Innovation and Entrepreneurship in Biotechnology, an International Perspective

Innovation and Entrepreneurship in Biotechnology, An International Perspective

Concepts, Theories and Cases

Damian Hine

Senior Lecturer, University of Queensland, Australia

John Kapeleris

Deputy CEO, Australian Institute for Commercialisation, Brisbane, Australia

Edward Elgar

Cheltenham, UK • Northampton, MA, USA

Published by
Edward Elgar Publishing Limited
Glensanda House
Montpellier Parade
Cheltenham
Glos GL50 1UA
UK

Edward Elgar Publishing, Inc.
136 West Street
Suite 202
Northampton
Massachusetts 01060
USA

A catalogue record for this book
is available from the British Library

Library of Congress Cataloging in Publication Data

Hine, Damian, 1962–
    Innovation and entrepreneurship in biotechnology, an international perspective : concepts, theories and cases / Damian Hine, John Kapeleris.
        p. cm.
    Includes bibliographical references.
    1. Entrepreneurship. 2. Biotechnology industries–Management. I. Kapeleris, John. II. Title.

    HD9999.B442H56 2006
    338.4′76606′0684–dc22   2005050163

ISBN-10: 1 84376 584 5
ISBN-13: 978 1 84376 584 4

Printed and bound in Great Britain by MPG Books Ltd, Bodmin, Cornwall

Contents

Acknowledgements

Damian would like to thank Maher Khaled for co-authoring Chapter 6 of this book and working closely with him in developing the important underlying concepts upon which the chapter is based. He would also like to thank Nicky Milsom for co-authoring Chapter 3 of the book. Special thanks for their input on both general and specific ideas go to Dr Ross Barnard, Mr Greg Orders and Dr Andrew Griffiths. Also thanks to Ms Micaela Buckley and Mr Paul Curwell for some excellent ideas that helped cement concepts in specific chapters.

John would like to thank his professional development mentors Professor David Wyatt and Mr Mel Bridges who have guided him through a successful career in the biotechnology industry. Thanks also go to Dr Craig Belcher for his support and motivation through the development of the book. John's learning and writing achievements would not have been fulfilled without the creative inspiration of Dr Ross Barnard, Dr Graeme Barnett and Mr Gerry Arkouzis.

The understanding and support of our families, though sometimes taken for granted, was absolutely critical to the successful completion of this book. For this Damian gives love and thanks to his wife Jo-Anne. John would like to thank his wife Sophie for the personal inspiration that is needed on a day-to-day basis when undertaking a significant project such as writing a book.

This book is dedicated to Amelia, Miranda, Isabella, Joanna and Peter.

Abbreviations

ACAP	Absorptive capacity
BBSRC	Biotechnology and Biological Sciences Research Council (UK)
BERD	Business expenditure on R&D
CDA	Confidentiality disclosure agreement
CI	Continuous improvement
COTS	Commercial off-the-shelf
CPM	Critical path method
CRO	Contract research organization
ELISA	Enzyme-linked immunosorbent assay
EMEA	European Medicines Evaluation Agency
FDA	Food and Drug Administration (USA)
GERD	Gross expenditure on R&D
GM	Genetically modified
HGP	Human Genome Project
IC	Intellectual capital
IP	Intellectual property
IPO	Initial public offering
LMO	Living modified organism
MoRST	Ministry of Research, Science and Technology (NZ)
MUCO	Multivalent coupling
NBF	New biotechnology firm
NDA	Non-disclosure agreement
NI	Neuraminidase inhibitor
NME	New molecular entity
NMR	Nuclear magnetic resonance
NPD	New product development
OLC	Organization life cycle
PACAP	Potential absorptive capacity
PCR	Polymerase chain reaction
PLC	Product life cycle
PM	Project management
PMA	Pharmaceutical Manufacturers Association
POC	Proof of concept

POLC	Population of organizations life cycle
RBV	Resource-based view (of the firm)
R&D	Research and development
SBDD	Structure based drug design
SBIR	Small Business Innovation Research (USA)
TCE	Transaction cost economics
VC	Venture capital

1. Introduction: innovation and entrepreneurship

INTRODUCTION – SCOPE OF THE BOOK

This book fills a market need in the fast-growing area of biotechnology management. Its key focus is the area of entrepreneurship and innovation in biotechnology, as it covers the main theoretical and practical aspects concerned with entrepreneurship in the biotechnology industry, focusing particularly on the innovation processes that underpin success for new biotechnology firms (NBFs). While the physical size of such companies may be small, resourcing and financing issues associated with major long-term R&D programmes are significant, as are potential returns. Entrepreneurship and innovation are major factors in all stages in the development of such companies; however it has been recognized that the biotechnology industry globally is lacking in managers and researchers with appropriate entrepreneurial and commercial skills (Batterham 2000; Sainsbury 2002).

The importance of the area is demonstrated in the priority afforded to entrepreneurship and innovation by universities, public sector agencies, governments and industry globally.

It is intended that this book will assist the reader to develop:

- The ability to differentiate between innovation and entrepreneurship in the biotechnology context;
- The ability to define the various forms of innovation present in the biotechnology industry;
- A detailed understanding of the strategic role of innovation, R&D and intellectual capital in biotechnology organizations;
- The ability to evaluate the impact of cycles on the competitiveness of biotechnology organizations and to make some predictions on these impacts;
- An ability to assess the external environment of biotechnology organizations from a scientific, regional, policy and resource perspective;
- The ability to define networks and alliances in the biotechnology industry, and assess their suitability in defined contexts;

- An appreciation of the impact and importance of globalization in the biotechnology industry and to biotechnology companies in terms of competitiveness and the imperative to innovate.

The book explores the role both entrepreneurship and innovation play in the competitiveness of biotechnology companies. Entrepreneurship is considered in terms of individuals/groups recognizing and acting upon opportunities in biotechnology-related markets. Innovation is viewed both as a creative process and as an essential element of an effective management structure and strategy for achieving growth in a biotechnology company. The book focuses on innovation and bioentrepreneurship in the biosciences side of the biotechnology industry rather than the more specified areas of agricultural or environmental biotechnology in order to maintain a balanced international perspective.

The book reflects two critical aspects for consideration of the entrepreneurial biotechnology company; the first with a company focus (internal environment), the second considers the position of the company in the biotechnology industry (external environment).

These two aspects highlight the importance of innovating to compete in the rapidly developing, technically sophisticated, biotechnology industry. The NBF must develop a market presence through an entrepreneurial orientation and innovative products, or research programmes, contractual services and internal processes, while avoiding operating in isolation.

It is the intent of the authors to reinforce the premise that entrepreneurship and innovation are key to achieving and maintaining competitiveness in the biotechnology industry. As in any industry, there are general management principles that apply alongside the specifics of the industry. These follow roughly an 80/20 rule; 80 per cent of management issues are generic across industries, but it is the 20 per cent that are unique to an industry that makes the difference. Knowledge of one of these components alone is insufficient; a combination of the two is required for successful management – in this case, in the biotechnology industry. This book combines both the generic and the specific management issues as much as possible, but its emphasis on matters of direct relevance to the biotechnology industry from an internal and external perspective is what makes it unique and distinguish it from other more generic texts on entrepreneurship and business.

Practical application to support the theoretical concepts is achieved through the many and varied case examples employed throughout the book.

The case studies and case examples take an international perspective. Such a perspective recognizes that best practice examples emanate not only from the USA where the biotechnology industry is more mature, but also from emergent industries in other countries. Much existing material originates in the USA and relies on the Unites States' unique (but not necessarily transferable) capital market structures and the free movement of activities from universities to the private sector supported by legislation such as the Bayh–Dole Act and the Stevenson–Wydler Act. The impact of measures such as the Bayh–Dole Act's 'use it or lose it' edict for university intellectual property (IP), are unique to the USA and are only under early consideration in most other countries. Although the USA does provide many examples of best practice due to the advanced nature of its biotechnology industry, each national industry environment is unique and lessons and practices are not necessarily directly transferable. There are a number of other examples of attempts to develop biotechnology policies nationally, for example South Korea and Singapore, which have instituted nationwide biotechnology policies, as well as publicly-funded support mechanisms. Singapore has expressly established biotechnology as one of its economic pillars (particularly when a previous pillar, the IT industry, collapsed). Cases are drawn from Australia, New Zealand, the UK, mainland Europe (including Germany, Denmark, Sweden, France and Switzerland), Asia (including Singapore, South Korea, India and Taiwan) and the USA, to highlight examples of different entrepreneurial biotechnology firms, and the industries in which they operate, providing a multiple perspective that avoids a single lens for analysis. The book therefore provides a very different perspective to conventional US material.

There are considered to be three distinct areas of biotechnology:

- Red – Biomedical, medical and human health (and animal health)
- Green – Agricultural biotechnology
- White – Industrial biotechnology

The emphasis of this book is on Red biotechnology, emphasizing human health, life-prolonging, life-preserving and life-improving technologies and developments. This in no way seeks to place Red biotechnology above Green or White; it simply displays the orientation of the authors' knowledge base.

INNOVATION

Innovation is a complex concept as there are a number of forms of innovation. The general definition is presented here followed by the more narrow definitions. Innovation in its widest sense is considered to be anything that is new to a business (Abernathy and Utterback 1978). The traditional concept of innovation is well documented and defined. In fact it has been defined, particularly for technological innovation, to such a degree that survey questions have been standardized in the OECD's Oslo Manual (1997, p. 31) which states, 'technological innovations comprise new products and processes and significant technological changes of products and processes. An innovation has been implemented if it has been introduced on the market (product innovation) or used within a production process (process innovation)'.

Innovation plays an important role in organizational and economic development, as evidenced by the large scope and sum of dedicated literature. It can be found in such areas as management (Damanpour 1991; Van de Ven 1986), learning (Cohen and Levinthal 1990; Nooteboom 1999), strategy (Lengnick-Hall 1992), clusters (Kenney and von Burg 1999; Pouder and St John 1996) and networks (Diez 2000; Robertson et al. 1996).

Furthermore, innovation has been studied from many disciplinary approaches, including economics (Baptista 2000; Karshenas and Stoneman 1993; Reinganum 1981), sociology (Woolgar et al. 1998; Rogers 1995), marketing and management (Dos Santos and Peffers 1998; Hannan and McDowell 1984; Robertson et al. 1996), geography and organizational ecology (Ciciotti et al. 1990; Antonelli 1989; Benvignati 1982; Oakey 1984; Thwaites and Oakey 1985).

Distinctions can be made between:

- Technological and non-technological Innovation, and
- Product and process Innovation (OECD 1992, pp. 27–9).

Technological versus Non-technological Innovation

Some authors (Damanpour 1991; Rogers 1995; Van de Ven 1986) have provided very broad definitions, describing innovations as new ideas or behaviours. It is the *perception* of newness, by those involved, which is central to Rogers' (1995, p. 11) definition of innovation. That is, an idea or item need not have scientific novelty to be regarded as an innovation (Schumpeter 1964, p. 59).

These definitions of innovation include both technological (products, services, processes) and non-technological innovations (organizational processes, administration systems) (Damanpour 1991, p. 556). For example, Schumpeter (1964, p. 59) describes five types of innovation:

1. Introduction of a new commodity
2. Introduction of a new production method
3. Opening up of a new market
4. Change in the source of supply
5. Re-organization of an industry.

The first two types are considered to be true technological innovation in the Schumpeterian sense, where technological innovations are new products and processes or significant changes of products and processes, and have an economic result. Thus, where improvements in quality and productivity outputs of a manufacturing process are required, the focus should be restricted to technological innovations.

Product versus Process Innovations

The OECD's *Oslo Manual* (1992; 1997) separates innovations into two types: product and process innovations. A product innovation is 'the commercialization of a technologically changed product'. Furthermore, a technologically changed product is one whose design is altered to provide new or improved performance over previous products. The product's characteristics, attributes, and design properties may be significantly changed – a major or radical product innovation – or an existing product may have significant performance improvements – an incremental product innovation.

A process innovation, alternatively, is a 'change in the technology of the production of an item [which] may involve new equipment, new management and organization methods, or both' (OECD 1997, p. 10). A process innovation can occur in any way a product is produced. In addition, a product innovation by one organization may be adopted as a process innovation by another organization. Process innovations can thus enhance quality and productivity outputs, and therefore provide significant competitive advantages.

Product innovation can take two broad forms:

- Substantially new products: we call this 'major product innovation'
- Performance improvements to existing products: we call this 'incremental product innovation' (OECD 1997, p. 29).

Major product innovation
This is a product whose intended use, performance characteristics, attributes, design properties or use of materials and components differs significantly compared with previously manufactured products. Such innovations can involve radically new technologies, or can be based on combining existing technologies in new uses.

Incremental product innovation
This is an existing product whose performance has been significantly enhanced or upgraded. This again can take two forms. A simple product may be improved (in terms of improved performance or lower cost) through

use of higher performance components or materials, or a complex product which consists of a number of integrated technical subsystems may be improved by partial changes to one of the subsystems.

In biotechnology and the pharmaceutical industry the approval of New Molecular Entities (NMEs) by the US Food and Drug Administration (FDA) are the most visible form of major product innovation. The FDA maintains its list of New Drug Approvals, however many of these approvals are for variants of existing drugs rather than entirely new drugs. Variants would be considered to be incremental product innovations. The FDA equivalent in Europe for drug approval is the European Medicines Evaluation Agency (EMEA).

As its name suggest the FDA is not concerned only with drugs. The products the FDA regulates include:

- Food – covering food-borne illness, nutrition, and dietary supplements
- Drugs – prescription, over-the-counter, generic drugs
- Medical devices – such as pacemakers, contact lenses, hearing aids
- Biologics – such as vaccines and blood products
- Animal feed and drugs – for livestock and pets
- Cosmetics – safety, labelling
- Radiation-emitting products – cell phones, lasers, microwaves
- Combination products – multi-category products which may include drugs and biologics as the move away from small molecule pharmaceuticals leads to more complex biologics in the drug development process.

Approval by the FDA does not assure success, nor does it assure safety, as has been recently demonstrated by Vioxx, the Merck product approved by the FDA in 1999 'for the reduction of signs and symptoms of osteoarthritis, as well as for acute pain in adults and for the treatment of primary dysmenorrhea' (Statement of Sandra Kweder, Deputy Director, Office of New Drugs, FDA, before the Committee on Finance, 18 November 2004).

The product is believed responsible for many thousands of heart attacks due to increased cardiovascular risk from the drug, particularly in the older target group of patients. Merck's share price plummeted over 25 per cent in one day on news of the withdrawal of Vioxx from the market due to side effects. However the FDA's web site provides solid reasoning for permitting a drug to market which has some risks attached:

> at the heart of all FDA's medical product evaluation decisions is a judgment about whether a new product's benefit to users outweighs its risks. No regulated product is totally risk free, so these judgments are important. The FDA will allow

a product to present more of a risk when its potential benefit is great – especially for products used to treat serious, life-threatening conditions (FDA Approvals for FDA-Regulated Products http://www.fda.gov/opacom/7approvl.html).

Of course this even displays the limits of the Schumpeterian concept of innovation. Vioxx reached the market; in fact it was very successful, having been approved in Europe, a number of countries in Asia, Australia, and New Zealand. It achieved commercialization and a high level of product diffusion through market up-take, no doubt assisted by Merck's extensive market and distribution network. Yet in the long run it will be a failure, as it has not achieved what the company had hoped it would in terms of risk mitigation while offering positive clinical results, and its commercialization ceased from September 2004. The legal battles, class actions, loss of reputation and market capitalization, as well as the marketing effort required for revival will probably result in a net loss from this product for Merck and will have major impacts on the industry.

The FDA has already had its reputation as the lead regulatory body in the world tarnished with Vioxx, Celebrex and a spate of recent product market withdrawals and warnings. Other major bodies have tended to follow most regulatory approvals handed down by the FDA, its pre-eminent position being based on strong clinical trials and stringent approval processes, to the point that Phase IV clinical trials are about to become de rigueur in the industry. Any flaw in the process will have major implications for the industry and will be felt far and wide. This itself will have a major impact on future product innovations, as it is likely to slow the already slow approval process, require higher significance and efficacy levels and reduce the number of candidates reaching the market.

Expect to see more stringency, longer approval lead times, compulsory Phase IV trials, post-market testing dramatically increased, separation of roles within the major approval bodies, major marketing campaigns to restore confidence in the system and life generally becoming more difficult for NBFs with products in the pipeline. This will be a significant story of not only innovation, but of diffusion. Much will be learnt from this and much will change as a result.

Process innovation
Process innovation refers to the adoption of new or significantly improved production methods. These methods may involve changes in equipment or production organization or both. The methods may be intended to produce new or improved products, which cannot be produced using conventional plants or production methods, or essentially to increase the production efficiency of existing products.

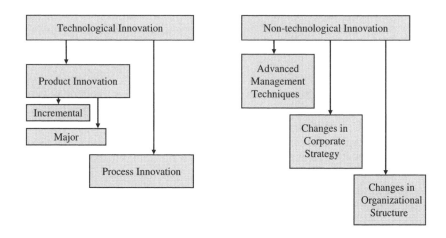

Adapted from ABS Innovation Survey 1995, 2003 and OECD Oslo Manual 1992

Figure 1.1   Forms of innovation

Later innovation survey guides differed from the OECD guidelines in some respects. The OECD's definition of innovation is restricted to technological innovation. The Australian Bureau of Statistics has widened this definition to include non-technological innovation. Non-technological innovation comprises three major aspects relating to the operation of a business rather than specifically to the products or services they produce. Figure 1.1 provides a summary of the major types of innovation referred to in this research study.

WHAT IS TRADITIONAL ENTREPRENEURSHIP?

The word 'entrepreneur' is derived from the French word '*entreprendre*' meaning 'to undertake' (Ronstadt 1985, p. 28). The traditional entrepreneur is one who undertakes to control, coordinate and assume the risk of a business in a competitive marketplace. Today's entrepreneurs possess those same features and have to be versatile in facing the challenges of a dynamic environment. Today's entrepreneur is an innovator and developer of ideas; he or she seizes opportunities and converts them into marketable entities; at the same time they have to lead a team, seek out capital and resources while creating something unique and of value to others (Montagno et al. 1986).

Although no single definition exists, it is commonly agreed that the entrepreneur is an agent of change, especially in a growing world of free enterprise and capitalism. In this sense, the traditional entrepreneur is the greatest risk-taker but ultimately claims the greatest rewards. This enviable nature of the entrepreneur spans every industry – there is one to be found in manufacturing, services, technology, agriculture and so on.

Entrepreneurship as Conceptual Orphan – Early Economic Writings

While the role of entrepreneurs and entrepreneurship has never been denied, its most debilitating facet is its limited, often cursory, coverage by the great thinkers on economics and business. John Kenneth Galbraith's eloquent tome, *A History of Economics: The Past As The Present* (1987), reflects that the mercantilists and their successors the entrepreneurs, being the *nouveau riche*, were not favoured by economists who, like the artists of the time, were patronized by the aristocracy and landed gentry, a class which felt threatened by the emergence of the Industrial Revolution. As Galbraith muses 'was the Industrial Revolution the product of inspired entrepreneurship? Was it an early step in a long process by which inventions, so far from being an independent innovating force, are the predictable achievement of those who, with brilliance and inspiration perceive the possibility of change' (1987, p. 58).

The great debates have considered groups whose ranks would largely consist of entrepreneurs; the mercantilists. Their identification often came in a political context in which landed aristocracy's pre-eminence was threatened by this 'new rich' class. The response came in the form of the intellectual elite in France at the time, the Physiocrats, led by Quesnay (1694–1774). Through their ideal of the *Produit Net* in which nature, land and its groups were considered the only source of wealth the Physiocrats sought to retain the landed aristocracy's rightful place at the head of the economic and social order, and negate the role played by the mercantilists, the entrepreneurs.

Even Adam Smith, despite being appointed for some time as Commissioner for Customs in Edinburgh, was no friend to the mercantilist, the merchant class. His distrust is evident in his passage, 'people of the same trade seldom meet together, even for merriment and diversion, but the conversation ends in a conspiracy against the public, or in some contrivance to raise prices' (Smith 1776 [1986]). The forerunner of the modern-day anti-trust, anti-competitiveness laws can be seen in these lines. This is despite strong belief that the greatest of all economic, and arguably social, steps forward occurred as a result of a combination of entrepreneurship and innovation.

It is not surprising then that the political agenda, which is always inextricably linked to economic writings, did not favour a positive portrayal of the role of the entrepreneur and of entrepreneurship. Where there is little support, or in the case of the Physiocrats, of Marx and to a lesser extent of

Smith, there is downright disdain, and tomes on the virtues and necessities of such groups will be difficult to obtain, unless by one of their own. It is this lack of consideration throughout the development of economic thought and principles, which has likely been the greatest contributor to the lack of a disciplinary base that entrepreneurship has experienced. There is also the fact that the small businesses created by entrepreneurs constituted a maligned area of the microeconomy until works such as Birch's (1979). Such events conspired to deny entrepreneurship its own lineage and body of works upon which to rely.

Thankfully for the discipline of entrepreneurship, the sentiment amongst great economic thinkers mellowed in the first half of the twentieth century. A better profile of an entrepreneur could not have been found than that envisaged by Weber as he explains what Schumpeter would probably refer to as creative destruction. Weber refers to the textile industry of the nineteenth century and its dramatic shift from the leisurely traditionalist business practice to that of entrepreneurial capitalism.

> . . . Now at some time this leisureliness was suddenly destroyed, and often entirely without any essential change in the form of organisation, such as the transition to a unified factory, to mechanical weaving etc. What happened was, on the contrary, often no more than this: some young man from one of the putting-out families went out into the country, carefully chose weavers for his employ, greatly increased the vigour of his supervision of their work and thus turned them from peasants into labourers. On the other hand he would begin to change his marketing methods by so far as possible going directly to the final customer, would take the details into his own hands, would personally solicit customers, visiting them every year, and above all would adapt the quality of the product directly to their needs and wishes. (Weber 1930, p. 125)

If this is not a scenario of a small business entrepreneur undertaking extensive process innovation, then what is?

For John Maynard Keynes, the main emphasis of his major work *The General Theory of Employment, Interest and Money* (1935), was stated succinctly by himself: 'The object of such a title is to contrast the character of my arguments and conclusions with those of the classical theory of [economics]' (p. 3). As Keynes himself indicates, 'The classical economists was a name invented by Marx to cover Ricardo and James Mill and their predecessors . . .' (p. 3).

So Keynes' emphasis was clearly upon developing a new economic thought for an era dominated by the Great Depression. Even with his eyes fixed squarely on economics Keynes in his General Theory still referred to entrepreneurs on numerous occasions, at times in some detail and usually in a favourable light. Just as Schumpeter believed unequivocally that innovation was an economic concept, Keynes saw entrepreneurship clearly as an economic phenomenon. This is based upon Keynes' distinction between speculation and enterprise. This distinction is underscored in Keynes' look at future policy directions in his concluding notes as he writes:

> Thus we might aim in practice at an increase in the volume of capital until it ceases to be scarce, so that the functionless investor will no longer receive a bonus; and at a scheme of direct taxation which allows the intelligence and determination and executive skill of the financier, the entrepreneur et hoc genus omne (who are certainly so fond of their craft that their labour could be obtained much cheaper than at present) to be harnessed to the service of the community on reasonable terms or reward. (1935, pp. 376–7)

We must accept that despite being written for a possibly different audience, and using some different terminology, many of the concepts that support our current research are available in the earlier writings, if we care to look. Further, as many of the writings contain anecdotal case studies, they are inherently flexible in their application.

Weber looks at entrepreneurship in his own current context and asks whether it differs from previous eras: 'At present under our individualistic political, legal, and economic institutions, with the forms of organisations and general structure which are peculiar to our economic order, this spirit of entrepreneurship might be understandable' (1930, p. 72). Is he writing in 1928 or 1998, without the reference the distinction is difficult to make?

Nevertheless those extolling the virtues of entrepreneurship remained limited in number and unsung in wider debates. Even in the post-war era other disciplines aligned with entrepreneurship, such as innovation, developed in leaps and bounds, assisted in no small part by the works of Joseph Schumpeter and other Austrian School economists such as Kirzner and Lachmann (though their disputes over the premise of equilibrium and disequilibrium remain, albeit posthumously). This impetus was added to by Rogers (1966), Abernathy and Utterback (1978), Tushman and Anderson (1986), and a succession of other subsequent writers. This provided the lineage, the chronology that all disciplines rely on in their necessity to establish their perceived centrality. The entrepreneurship lineage, the extant chronology of important major works is a relatively short one.

As a result, entrepreneurship has been bereft of dominant paradigms, it has hence had to borrow from other disciplines to build its theoretical depth and undertake its analyses. Major borrowed concepts upon which economic entrepreneurship has been based in recent years include

- Transaction cost economics (TCE)
- Resource-based view of the firm (RBV)

- Institutional theory
- Networks and alliances.

These concepts will be raised and explored throughout the ensuing chapters as their influence and value is undeniable in explaining entrepreneurial phenomena.

ENTREPRENEURSHIP AND INNOVATION

Post-Fordist or post-industrial influences (Keeble 1997) have created a less antagonistic, even supportive environment for high technology companies and entire industries to emerge and grow. Such influences include rapidly changing environments, moves away from mass production toward customization and mass-customization, the application of generic technology which advances communication and production processes, a declining emphasis on price competition and price competitiveness through volume output and economics of scale. The central theme of this development has been the appropriate application of technology through innovation, which itself is dependent upon the skills and risk taking of an entrepreneur. Put simplistically in Figure 1.2, the essential features have a strong causal relationship.

These post-industrial influences have created an environment conducive to the growth of high tech firms and entire industries. These influences include:

- Rapidly changing environments, particularly globalization impacts;
- A trend from mass production toward customization;
- The application of generic technology that advances communication and production processes;

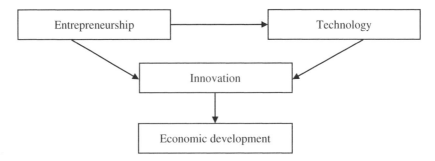

Figure 1.2   The relationship between factors in the improved standing of entrepreneurial businesses

- A declining emphasis on price competition and price competitiveness through volume output and economics of scale.

The central theme of this development has been the appropriate application of technology through innovation, which itself is dependent upon the skills and risk taking of an entrepreneur. Simplistically, the essential features of both entrepreneurship and innovation have a strong relationship; it is the causality that is debated.

In this situation, the fundamental role of innovation is to achieve profitability for the firm; it is a means to an end, not an end in itself. While for the individual firm the goal is profitability, the collective impact of this goal is economic change. Governments around the world therefore regard biotechnology as important since it has the potential to bring economic prosperity as well as ushering in a new economic order in the national/regional economy, creating a shift away from dependence on commodities and traditional industries for revenue.

The essential form of entrepreneurship is considered to be the 'creation of new enterprise'. Therefore an 'entrepreneur' could be considered to be one who creates the new enterprise. It has also been identified that 'entrepreneurship' is a 'process of becoming rather than a state of being' (Mazzarol et al. 1999). This means that small businesses often commence as self-employed people or micro-businesses. If their existence is justified then they are provided the opportunity to grow and create employment for others. Innovation is an essential element of the entrepreneurial effort. Entrepreneurs must innovate in order to be successful. The form this innovation takes can vary according to the type of entrepreneur, the market and product requirements and the life cycle stage the business is currently experiencing. The type of entrepreneur is an important element in the success of a biotechnology company. Although there is no archetypical entrepreneur, there are certain recognizable characteristics of entrepreneurs that are worth noting.

There are three things that according to Schumpeter (1934, p. 89) make it easier to start a new enterprise:

- Offering goods and services that are new
- Some new enterprises use new methods of production or marketing or draw on new sources of materials or components to offer more cheaply or conveniently than competitors
- Some new enterprises are able to establish a new industry structure or exploit a new style of regulatory environment more effectively and/or rapidly than their rivals.

It is worth bearing in mind, the impact on the economy if, instead of the usual 10 to 15 per cent of small business owners being entrepreneurs, 90 per cent of them were.

There is also an important distinction to be made between the terms 'innovation management' and 'entrepreneurship'. One is managing the research and particularly the development process, the other is a far more broad strategic approach to running a business.

Keynesian economic principles, as robust as they have been in the past, were written in an industrial era, in a time before technology changed the imperative from size and scale to flexibility. The sector of the economy to benefit the most from these revolutionary developments has been that sector which already possessed many of the features which would provide competitive advantages: flexibility, willingness to incorporate new approaches and technologies, low capital base, responsiveness to customer needs, i.e. small business.

There was one feature which was generally lacking in small businesses which in the Keynesian era had precluded them from competitiveness in many established industries and led them to be written off as an anachronism by many of the great economic thinkers of history, such as Keynes himself, Schumpeter, Marx and Smith. They lacked knowledge of the market in which they operated. With small production runs, lacking in resources due to a low capital base, little market share and hence negligible market power, small firms found it difficult to access the necessary infrastructure they required for competitiveness. Competitiveness could only be obtained through reducing the cost of utilizing the available industrial infrastructure – namely roads, rail, ports and more recently airports, as well as the education system, most efficiently. The only way to achieve efficiency was through reducing the per-unit costs of transporting products to market. Hence the reliance on economies of scale. Those firms which had achieved economies of scale had access to the available infrastructure and therefore had an advantage over those that did not. Not surprisingly then small business was not competitive in markets which had become established and where size was the key to competitiveness.

With size, came access to not only the physical infrastructure but to other advantageous factors such as education, and information, each of which could be restricted under the traditional industrial infrastructure to those firms who had the means to access. This meant that market knowledge lay in the hands of large firms, thereby improving their ability to make decisions based upon reliable data.

**Technology Development and Entrepreneurship and the Demise
of Traditional Microeconomic Thinking**

In the years since 1985 there has been an increasingly viable body of literature emerging which has supported the focus on small business as the engine room of economic activity in national economies. Much of this literature has unfortunately flown in the face of traditional economic principles and has therefore been severely critiqued as lacking theoretical underpinnings. The emergence of small business as a force within industrialized economies has coincided with the failure of Keynesian economic principles to be effectively implemented. This can largely be traced to the oil crisis of the early 1970s and to the accelerated development of technologies which improved the productivity of manufacturing and reduced the reliance upon traditional petroleum products.

This emerging technology has been significantly supported by corresponding developments in computerization and transistorization, initially specifically in the electronics-based industries and then diffusing to other related and then unrelated industries. The uptake of the technological advancements, particularly in manufacturing has meant that production costs have been dramatically reduced for many goods and indeed for services.

The impact of this burgeoning technology base has been twofold: It has led to the reduction in the costs of production and subsequently the price of products which come to market. This has meant that firms no longer need to rely upon as large a capital outlay as previously to achieve sufficient volume of production and therefore competitiveness in a market.

The reduction in the cost of production and the subsequent lowering of prices for products has meant that many products which were previously seen as luxury items have become more available to a larger proportion of the population. This has led to a reduced preference for mass produced products and an increase in the desire for customized items. Such a demand change has run counter to the traditional economies of scale requirement for competitiveness. Given the emerging technologies which are diffusing so rapidly as to impact on the majority of secondary and tertiary industries, the markets in which firms are operating are also changing rapidly. In parallel the technology itself allows for increased flexibility in the production process and an improved ability for those firms most willing to incorporate those technologies, the innovative firms, to meet the changes in demand rapidly. For instance retooling in many industries is not such a major or time and resource dependent task as previously. Hence the requirement for large production runs to reduce costs of production have been reduced. This again parallels the changing tastes and preferences of consumers toward customized over mass-produced products.

The consequence is that economies of scale no longer hold the key to competitive advantage as they previously did. Flexibility, in production and in meeting the rapidly changing demands of consumers is increasingly the order of the day.

In Chapter 2 a more thorough discussion of the emergence of entrepreneurship in the biotechnology context is presented. The generic material in this chapter, coupled with the biotechnology specific exploration of entrepreneurship in Chapter 2 will set the scene for the detailed analyses and discussions in the ensuing chapters.

REFERENCES

Abernathy, W.J. and J.M. Utterback (1978), 'Patterns of industrial innovation', *Technology Review*, **80**, June–July, 2–9.

Antonelli, C. (1989), 'The role of technological expectations in a mixed model of international diffusion of process innovations: the case of open-end spinning rotors', *Research Policy*, **18** (5), 273–88.

Australian Bureau of Statistics (1995), *The Innovation Survey*, Canberra: AGPS.

Australian Bureau of Statistics (2003), *The Innovation Survey*, Canberra: ABS.

Baptista, R. (2000), 'Do innovations diffuse faster within geographical clusters?', *International Journal of Industrial Organisation*, no.18, 515–35.

Batterham, R. (2000), *The Chance to Change*, Chief Scientist of Australia, Canberra, DISR.

Benvignati, A. (1982), 'Interfirm adoption of capital-goods innovations', *Review of Economics and Statistics*, **64** (2), 330–35.

Birch, D. (1979), *The Job Generation Process*, Cambridge, MA: MIT Study on Neighborhood and Regional Change.

Ciciotti, E., N. Alderman and A. Thwaites (1990), *Technological Change in a Spatial Context: Theory, Empirical Evidence and Policy*, Berlin: Springer-Verlag.

Cohen, W.M. and D.A. Levinthal (1990), 'Absorptive capacity: a new perspective on learning and innovation', *Administrative Science Quarterly*, **35**, 128–52.

Damanpour, F. (1991), 'Organizational innovation: a meta-analysis of effects of determinants and moderators', *Academy of Management Journal*, **34** (3), 555–90.

Diez, J. (2000), 'The importance of public research institutes in innovative networks – empirical results from the Metropolitan Innovation Systems Barcelona, Stockholm and Vienna', *European Planning Studies*, **8** (4), 451–63.

Dos Santos, B. and K. Peffers (1998), 'Competitor and vendor influence on the adoption of innovative applications in electronic commerce', *Information & Management*, **34** (3), 175–84.

FDA (2004), 'Approvals for FDA-Regulated Products', Available at: http://www.fda.gov/opacom/7approvl.html.

Galbraith, J.K. (1987), *The History of Economics: The Past as the Present*, London: Penguin Books.

Hannan, T. and J. McDowell (1984), 'The determinants of technology adoption: the case of the banking firm', *Rand Journal of Economics*, **15** (3), 328–35.

Karshenas, M., and P. Stoneman (1993), 'Rank, stock, order, and epidemic effects in the diffusion of new process technologies: an empirical model', *Rand Journal of Economics*, **24** (4), 503–28.

Keeble, D. (1997), 'Small firms, innovation and regional development in Britain in the 1990s', *Regional Studies*, May, **31** (3), 281–94.

Kenney, M. and U. von Burg (1999), 'Technology and path dependence: the divergence between Silicon Valley and Route 128', *Industrial and Corporate Change*, **8** (1), 67–103.

Keynes, J.M. (1935), *The General Theory of Employment Interest and Money*, New York: Harbinger.

Kweder, Sandra (2004), Statement before the Committee on Finance, 18 November 2004, by Sandra Kweder, Deputy Director, Office of New Drugs, FDA, Approvals for FDA-Regulated Products, Available: http://www.fda.gov/opacom/7approvl.html.

Lengnick-Hall, C. (1992), 'Strategic configurations and designs for corporate entrepreneurship: exploring the relationship between cohesiveness and performance', *Journal of Engineering and Technology Management*, **9** (2), 127–54.

Mazzarol, T., T. Vollery, N. Doss and V. Thien (1999), 'Factors influencing small business start-ups. A comparison with previous research', *International Journal of Entrepreneurial Behaviour and Research*, **5**(2), 48–62.

Montagno, R., D. Kuratko, and J. Scarcella (1986), 'Perception of entrepreneurial success characteristics', *American Journal of Small Business*, **10** (3), 25–46.

Nooteboom, B. (1999), 'Innovation and inter-firm linkages: new implications for policy', *Research Policy: Amsterdam*, **28** (8), 793–808.

Oakey, R. (1984), *High Technology Small Firms: Regional Development in Britain and the United States*, London: Frances Pinter.

OECD (1992), *OECD Proposed Guidelines for Collecting and Interpreting Technological Innovation, Oslo Manual*, 3rd edn, Paris, OECD.

OECD (1997), *OECD Proposed Guidelines for Collecting and Interpreting Technological Innovation, Oslo Manual*, 2nd edn, Paris, OECD.

Pouder, R. and C. St John (1996), 'Hot spots and blind spots: geographical clusters of firms and innovation', *Academy of Management Review*, **21** (4), 1192–225.

Reinganum, J. (1981), 'Market structure and the diffusion of new technology', *Bell Journal of Economics*, **12** (2), 618–24.

Robertson, M., J. Swan and S. Newell (1996), 'The role of networks in the diffusion of technological innovation', *Journal of Management Studies*, **33**(3), 333–59.

Rogers, E. (1966), *Diffusion of Innovation*, 2nd edn, New York: The Free Press, Collies-Macmillan Ltd.

Rogers, E.M. (1983), *Diffusion of Innovations*, 3rd edn, New York: Free Press (Cited in M. Robertson, J. Swan, and S. Newell, (1996), 'The role of networks in the diffusion of technological innovation', *Journal of Management Studies*, **33** (3), 333–59.)

Rogers, E.M. (1995), *Diffusion Of Innovations*, 4th edn, New York: Free Press.

Ronstadt, R. (1985), *Entrepreneurship: Text, Cases and Notes*, Massachusetts: Lord Publishing.

Sainsbury, Lord, (2002), 'Strong government support for UK biotech', *Biotechnology Investors' Forum Worldwide*, **2**, 20–22.

Schumpeter, J.A. (1934), *The Theory of Economic Development*, Cambridge, MA: Harvard University Press.

Schumpeter, J. (1964), *Business Cycles: A Theoretical, Historical, and Statistical Analysis of the Capitalist Process*, New York: McGraw Hill.

Smith, A. (1986), *The Wealth of Nations*, Harmondsworth: Penguin. Originally published in 1776.

Thwaites, A. and R. Oakey, (1985), *The Regional Economic Impact of Technological Change*, London, Frances Pinter.

Tushman, M.L. and P. Anderson (1986), 'Technological discontinuities and organisational environments', *Administrative Science Quarterly*, **31** (3), 439–65.

Van de Ven, A. (1986), 'Central problems in the management of innovation', Management *Science*, **32** (5), 590–607.

Weber, M. (1930), *The Protestant Ethic and the Spirit of Capitalism*, New York: Scribner.

Woolgar, S., J. Vaux, P. Gomes, J. Ezingeard and R. Grieve (1998), 'Abilities and competencies required, particularly by small firms, to identify and acquire new technology', *Technovation*, **17** (8/9), 575–84.

2. Entrepreneurship in the biotechnology context

INTRODUCTION – THE BIOTECHNOLOGY INDUSTRY AT A GLANCE

The biotechnology industry is a science-led industry with R&D cycles and product lead times up to 20–30 years duration (Champion 2001). In the 1950s and 1960s, the biotechnology industry was restricted to relatively established activities in the food and beverage, primary and secondary metabolite fermentations, and waste treatment industries (Barnard et al. 2003).

As the expansion in the basic science underpinning modern biotechnology occurred, newer developments in microbial genetics, enzyme and cellular technologies, and bioremediation drove the development of the science in the industry. With the ongoing development of recombinant DNA techniques, opportunities in both the science and the commercialization of the scientific research blossomed.

From the late 1970s and early 1980s, the USA led the world in the rapid formation of start-up companies specializing in the 'new' biotechnologies. While most other countries lagged behind the USA both in the number and size of such start-up companies, many European countries have nevertheless seen considerable activity in this sector, with a dramatic increase in companies achieving initial public offering (IPO) and a greater number of privately-held companies. These companies now cover the full spectrum of what is regarded today as biotechnology; from genomics, proteomics, and bioinfomatics companies, through to biopharmaceutical, probiotic, and environmental biotechnology companies.

'Biotechnology' by accepted definition, involves the use of living organisms or parts of living organisms through biological processing to develop new products or provide new methods of production. Biotechnology as a discipline is very old, dating back to the first production of beer, bread and wine, using yeast as the living organism. However, significant advances have been achieved through modern biotechnology, which dates back to the 1950s with the elucidation of the structure of DNA by Watson and Crick.

Biotechnology covers a diverse range of fields, including medicine, therapeutics, agriculture, food processing and environmental maintenance.

Biotechnology is an emerging and fast-growing industry that has the potential to revolutionize society in the next decade. It promises to provide improved healthcare and treatment of disease, new therapeutics, improved diagnostics, increased yield in food production and improved environmental management.

The major technology areas in biotechnology and examples of their application are summarized in Table 2.1 (BIO, 2001).

The biotechnology industry is characterized as follows:

- Medium to very long product development lead times;
- Capital-intensive;
- Highly regulated;
- Extensive skill sets and technical knowledge required;
- One of the most research-intensive industries in the world;
- In many cases ethical clearance is essential, especially for any animal/human testing;
- Intellectual property protection is an essential element of success for most biotechnology companies;
- Strong linkages and strategic alliances established with universities, institutions and other biotechnology companies;
- Capital raising is essential and consumes a significant amount of time and resources throughout an organization's life cycle.

Many biotechnology companies specialise in activities related to their expertise or core competencies. For example, a small biotechnology company may be involved in drug discovery focusing on the identification of new drug targets. However, they may not have the necessary capital resources or experience to take a particular drug candidate to market. It can take at least US$100 million to do this (Woicheshyn and Hartel 1996). A biotechnology company therefore will require a strategic alliance with a large pharmaceutical company to take the product to market. The process from target identification to final market launch may take up to 16 years (Woicheshyn and Hartel 1996; Ernst & Young 2000). Another biotechnology company, for example, may be focused on applied research where value is generated through the development of intellectual property, which subsequently may be licensed or sold to a larger biotechnology or pharmaceutical company.

Globally the biotechnology industry is shifting its value creation emphasis from R&D to manufacture and marketing as more products emerge from the R&D pipeline and the industry as a whole matures (Ernst and Young 2003). Yet, due to such factors as industry concentration, resource access, sophistication of financial markets and quality of science, the industry has

Table 2.1   Major technology areas and their application in biotechnology

Technology	Applications
Monoclonal Antibody Technology	● Diagnose infectious diseases in humans, animals and plants ● Detect harmful micro-organisms in food ● Distinguish cancer cells from normal cells ● Locate environmental pollutants
Cell Culture Technology	● In-vitro propagation of micro-organisms ● Embryonic stem cells and tissue replacement therapies ● Replace animal testing for evaluating safety and efficacy of medicines
Cloning Technology ● *Molecular cloning* ● *Cellular cloning* ● *Animal Cloning*	● Recombinant DNA technology ● Study genetic diseases ● New drug discovery ● Regeneration of transgenic plants from a single cell ● Generation of therapeutic cells and tissues ● Pharmaceutical manufacturing ● Improved animal herds
Genetic Modification Technology	● New medicines and vaccines ● Treat genetic diseases ● Enhance bio-control agents in agriculture ● Increase yields in agriculture ● Improve nutritional value of food ● Decrease water and air pollution ● Develop biodegradable plastics
Protein Engineering	● Development of new and improved proteins ● Development of new biocatalysts (enzymes) ● Cleaner more efficient production using biocatalysts instead of chemical catalysts
Hybrid Technologies ● *Biochip Technology* ● *Biosensor Technology*	● Genetic analysis- genetic diseases ● Multi-array detection of infectious diseases ● Detect environmental pollutants

Table 2.1　(continued)

Technology	Applications
• *Tissue Engineering* • *Bioinformatics*	• Point of care bedside testing of patients • Measure freshness and safety of food • Biodegradable tissue scaffolding • Tissue therapies – skin, cartilage etc • Replacement tissues through cell culture • Map genomes and identify genes • Determine protein structures • New therapeutic targets • Analyse gene mutations

Source:　Adapted from the Biotechnology Industry Organization (BIO) Report, 2001

developed at a differential pace in different nations. In evolutionary stage Australia trails well behind the USA and Canada. The UK, through the Biotechnology and Biological Sciences Research Council (BBSRC), has built a strong biotechnology profile in a relatively short space of time, but based on a long history of fundamental biological research, epitomized by the discovery of the structure of DNA and pioneering work on determination of the three-dimensional structure of proteins in the 1950s.

This emerging biotechnology industry has become a major engine for economic growth in knowledge-based economies. The model that has been proven to be economically successful in many countries is to have clusters of biotech companies situated around major publicly-funded research universities and institutes. In such a model, there is a tight linkage not only between the research in the public sector fuelling the private sector but also with the supply of new staff with the training relevant to the growing biotech companies.

Biotechnology has been recognized worldwide as a critical sector to national economies. However investment is crucial to the growth and continuity of the industry, as it is an industry characterized by high-cost research and development (R&D), limited commercialization, and rapid change brought about by constant technological developments and scientific advances.

There is an imperative for every country with a fledgling biotechnology industry to seek to build a platform for rapid growth and expansion of their industry. Just how to achieve this is a perplexing issue. Market-driven boom–bust cycles evident in other high-tech industries also impact on the biotechnology industry disregarding the obvious tangibility of the product and the depth of the science backing the industry. Two key characteristics can be used to differentiate the structural characteristics of the biotechnology industry from other high technology sectors (Powell 1998). These are: diffusion of innovations and practices, and innovation speed. These two important associated features will be discussed in depth in Chapter 7.

COMPETING FORCES IN BIOTECHNOLOGY MARKETS

In many industries, technological developments have increased the diffusion of innovations through improved speed and quality of communication, improving market knowledge of both producers and consumers. The Internet, e-mail, access to generic technologies (such as commercial off-the-shelf technology), have all contributed to the speed at which ideas and products spread. However the main issue is that for biotechnology, due to high capital costs, technology diffusion will be slow without significant levels of investment. This investment tends to come either from large, established pharmaceutical companies (Big Pharma) or from public and private investors. As opposed to the IT industry where $2–3 million would be required to get software designed, developed and on to the market within a six to 12-month period; in biotechnology, 'a typical drug takes 15 years and $500M to bring to market' (Champion 2001).

Furthermore, there is a very distinct difference between basic/blue sky research, which is problem-solving research for the benefit of health, largely undertaken by public organizations such as universities, government agencies and research institutes, and incremental/applied research and development which private sector firms are more likely to be involved in. Big Pharma are increasingly outsourcing their R&D (an increase from 5 per cent to 25 per cent since 1995), as they merge and seek to concentrate on market sales (Andrews 2002). Such a trend indicates that these companies recognize the long-term viability and importance of the research and diffusion of innovation process to new product development. A point which is missed by short-term, market-driven time horizons which are at odds with these industry structural characteristics.

Product diffusion is a problematic area for many biotech firms. Realistically most are research-oriented firms surviving on their inventiveness and innovativeness, based upon the quality of the science and originality of the research, usually indicated by the extent and value of intellectual property such as patents. The manufacturing environment is not a facet of business most are adept in. In fact, many small biotech firms, the type seeking to list or recently listed, choose to stick to the R&D and license their IP to other manufacturers for production. This allows the biotech firm to focus on their core business, R&D, particularly as such licensing deals can be lucrative. By way of example, Genetic Technologies (GTG), a small biotech company, in 2002 licensed its non-coding DNA patents to Myriad Genetics. An initial payment of $1.85 million was paid to GTG permitting it to expand its research agenda (Trudinger 2002).

For the larger and even many medium biotech firms this is not really an option. Not only does the firm invent products, it must also bring them to market to ensure the cash flow that will fund the 'innovation stream'. The dilemma is first for firms to work out their strategic position and where their core business lies along the value chain. If they are more than an R&D firm then they must manage product development and product diffusion effectively. This therefore requires significant capital resourcing and under-standing by investors of the long time horizons involved in the industry. If expectations are for rapid growth and return on investment, as in IT, then these expectations will not be met, funding will dry up, further slowing the diffusion of innovation process.

It would be dangerous to withdraw investment funding at this stage where, more than ever before, it can be argued that the industry is on the brink of making a commercial return for investors. However, the short-term demand on returns made by market-driven policies and practices prove to be disruptive rather than enhancing the long-term sustainability of the industry. While market capitalization is important to this industry, so too is its scientific base. Market-driven cycles can only serve to exacer-bate a split between biotechnology and its scientific base – where capabil-ity activities should be seeking to build this (Hine and Griffiths 2004).

Further, financial benchmarks are critically important as a basis for judging investment decisions for listed companies. Yet how can we have a price to earnings ratio for a company, which has no earnings? Put simply many of the biotech ventures don't have sales revenues. Down the track when their licenses start producing progress payments there will be some-thing to consider. Also where the venture undertakes such commercial tasks as contract research there will be revenue to consider.

The forces impinging on the growth and development of these two indus-tries can be loosely located into two camps. The scientific and technical forces often associated with technology push innovation from scientists and researchers in biotechnology and technical specialists in IT. Evidence of performance and development in this area is indicated by the intellectual property which is an outcome of their efforts. In biotechnology this is most clearly evidenced through two means: patents and publications. In IT the evidence is more difficult to pinpoint. For information technology, patents tell part of the IP story, though copyright and other IP protection techniques are employed.

The market forces and the desire to commercialize ideas and research through product innovation, providing a revenue stream and a return to investors also has a major influence on the development of these industries.

Each will have a differential impact on the level and type of development occurring within an industry. These two forces are often treated as competing explanations for growth of industries. For instance the market forces approach embodies the market as the dominant originator of influences on development. In contrast, the scientific and technical forces approach focuses on the intellectual property and intellectual capital creating the opportunities which are simply acted upon by business and industry. Neither is superior, preferred or mutually exclusive of the other. They can be, however, two competing forces in industry development. However when the forces are harmonized, then the influence on industry development can be positive. These perspectives are analogous explanations of how high technology industries may develop. In this sense the two forces provide parallax view of the phenomenon. Integrating these two perspectives is essential for gaining a full appreciation of the creation of sustainable high technology industry practices.

ENTREPRENEURSHIP IN BIOTECHNOLOGY

Fundamentally, entrepreneurship is considered to be the 'creation of new enterprise'. It incorporates both entrepreneurs as individuals and entrepreneurial firms. The entrepreneur is therefore the individual who creates the new enterprise. In the context of biotechnology, the bioentrepreneur operates in a knowledge-based and science-based industry, where competitive advantage is achieved through the effective management of intellectual property emanating from good science. The bioentrepreneur is often a scientist/researcher-turned-entrepreneur who wishes to see their research successes put into practice through commercialization. In Chapter 1, the essence of entrepreneurship has already been stated as a process of becoming rather than a state of being (Mazzarol et al. 2000). This observation is also appropriate for the bioentrepreneur, as they go through a metamorphosis from researcher to fully-fledged entrepreneur.

The following essential elements of entrepreneurship apply in the biotechnology industry. These include:

- Opportunity recognition
- Market knowledge

- Product knowledge
- The desire and ability to innovate
- The ability to assemble and effectively employ resources
- A propensity for calculated risk-taking which is the most recogniz-
  able of all elements amongst entrepreneurs.

It is argued that the difference between entrepreneurship in biotechnology and other industries is not in the essential ingredients (which are largely
the same), but in their proportions (which will vary). This is illustrated
by comparison between the biotechnology and IT industries, the most
prominent of the high-technology industries. This analysis both compares
and contrasts these two industries which are often considered in tandem,
while their antecedents, modus operandi and structure differ dramatically.

Table 2.2 outlines a comparison between Biotechnology and IT showing
how the proportions may differ. It is important to make this distinction
between bioentrepreneurship and entrepreneurship in other industries,
as exemplified by the IT example. While the technopreneur exists in all
high-technology industries, the bioentrepreneur is a further subset of this
genus.

Table 2.2   Differentiating features of IT and biotechnology

Features	Information Technology (IT)	Biotechnology
Product life cyles	Very short – often 6–12 months	Medium to long
Technological requirements	While a university degree is available, most IT skills are learnt by doing	In developing new products, extensive skills sets and technical knowledge are required. PhDs are common
Resource requirements	Limited to labour costs, hardware, software and overheads	This is a capital-intensive industry with extensive sunk costs
Capital raising	Most start-ups proceed to initial product launch without backing	Capital raising is essential and occupies much of the resources and energy of the bioentrepreneur
R&D cycle	Quite short – particularly where virtual teams collaborate Moore's Law	These vary – for incremental developments they can be one to two years – for discovery developments they can be 20–30 years

Table 2.2   (continued)

Features	Information Technology (IT)	Biotechnology
Regulatory requirements	The market moves so rapidly that regulation does not keep up. Corporate governance is largely internal	Extensive regulation which severely impacts on the product development process – this is a highly regulated industry
Ethics	Ethical issues do not predominate in this industry. Codes exist but are often voluntary	Ethical clearance is essential for any animal or human testing. Ethics in biotechnology is a public issue
Extent of R&D	Linked to the R&D cycle. R&D on individual products is not extensive but many products can be concurrently developed	The trial process – animal, chemical and field trials provide evidence that this is a long intensive process
Linkages	Collaboration is often between companies and individuals. Smaller firms innovate and then look to sell either their product or the business to larger firms	Linkages usually include institutions such as universities, CSIRO or multinationals. Small firms innovate and are often acquired by larger firms
Product development and product launch	Product development continues well after product launch (much to our annoyance)	Due to ethical and regulatory issues product development must be completed prior to launch
Intellectual property	Few patents, some trademarks, design copyright most prevalent. Intellectual property control difficult to maintain	Patents most prevalent, some trademarks. IP control is essential to the success of most companies. It is also a substantial financial burden

TECHNOLOGY AND OPPORTUNITY IN BIOTECHNOLOGY

It is hardly surprising that with technology fundamentally incorporated into its meaning, biotechnology opportunities are very dependent on technology. While the science is critically important, it was only due to earlier discoveries by such luminaries as Linus Pauling's application of electrophoresis to separate proteins in 1949, Hershey and Chase's use of radioactively labelled E. coli to support DNA as the carrier of heredity in 1950, and Wilkins and Franklin's use of crystallography and X-ray diffraction to determine the DNA outline in 1951 that Watson and Crick were able to achieve the discoveries they did. Cohen and Boyer, while fortuitously meeting at a conference in Hawaii and realizing they could assist each other's research projects, could not have achieved their breakthrough in recombinant DNA without a number of advances including: advances in spectrometry and magnetic resonance, building upon the work of Jacob and Monod's propositions on cell structure and gene expression; Nirenburg's cracking of the genetic code in 1964; isolation of restriction enzymes in 1970 and Berg's synthesis of the first recombinant DNA molecule in a test tube in 1972. This is not to detract at all from these pivotal achievements, it is simply a recognition that these scientists were building upon and benefiting from advanced equipment and newly developed scientific techniques which enabled these great discoveries to be made. Further developments since 1973 such as Mullis' invention of the polymerase chain reaction (PCR) in 1983 has meant that DNA fragmenting and cloning is now rapid and stable (though the diffusion of this technology was slowed by its high establishment costs).

The chronology of the biotechnology industry (as opposed to the science) from Genentech's establishment of Boyer and Swanson in 1976, has also been a chronology of dramatic technological development to the point that PCR, PAGE, mass-spectrometry, NMR, electron crystallography and many other developments underpin the advances in science. Hence the opportunities created in science and for business are opportunities created by technology. This technology can be considered in the sense envisaged by Woodward in the 1950s, as the combination of man-made equipment and the people who use it, rather than the equipment alone.

The important point to be made here is that entrepreneurship is about opportunity recognition and then realizing these opportunities. It is the dramatic advances in technology, paralleled by scientific discovery, which have created the footing for an industry. Entrepreneurs have populated this industry after the scientists and the technicians. Because of the technical and scientific understanding required to operate in many parts of the industry, entrepreneurs have not been attracted in large numbers from other industries (as is often posited in micro-economics). Hence many of the entrepreneurs have been home grown: research leaders/star scientists who have taken up the challenge themselves, either by choice or necessity, to progress their intellectual property along a commercialization path.

While the biotechnology industry has not had the benefit of an influx of entrepreneurs from outside, it has been fortunate to co-exist with its big brother, the pharmaceutical industry. The benchmarks and knowledge gained from the centuries-old pharmaceutical industry have been both a boon and a poison chalice.

New industries invariably emerge from established industries. This creates its own dilemmas of whether to follow or ignore the culture of its parent, establishing industry standards, garnering resources from the parent, building on existing technology or instigating new technology, relying on human capital from the parent or creating a new human capital base. For biotechnology there was an advantage. By its own cross-disciplinary nature it is recombinant. This allows it to break free from many of the constraints which have hindered the pharmaceutical industry of declining R&D spending, industry standards that stifle innovation, high levels of industry concentration leading to a concentration on building market power through scale rather than innovation. Of course there is the fundamental element underpinning the pharmaceutical industry which biotechnology has been unable to escape – the patent system. This will be a topic of discussion in Chapter 6.

Just as technologies have been enabling, and at times disabling, so have other features of the biotechnology industry. This is where this industry differs little from others. Yet it does differ in its current and potential size, extensive scientific and theoretical basis and its likely sustainability. While it has already suffered the pains of a 'tech bubble', it has recovered already more positively and has already usurped IT as the highest growth industry globally.

The industry has also developed so rapidly that it is spawning its own new industries in nanotechnology (though that has been recognized as an industry since the 1950s) and bioinformatics, as well as the various 'omics' such as genomics, proteomics, nanomics, metabolomics, pharmacogenomics.

WHAT IS BIOENTREPRENEURSHIP?

Bioentrepreneurship is a relatively recent term outside the USA. The USA is often considered to be the motherland of biotechnology, though many believe that given Watson and Crick's contribution to the science underpinning biotechnology, the UK deserves that status. Nevertheless what is less disputed is the birthplace of bioentrepreneurship. The USA is certainly the place where the bioentrepreneur has operated for over three decades. Before the IT giants of Microsoft, Sun, and Cisco were a twinkle in Steve Burrell's eye, Genetech and Amgen were making headway expanding through their own research programmes, competing and in critical cases collaborating with the major established pharmaceutical companies known as Big Pharma. These first biotechnology companies were populated by scientists, assisted by merchant bankers and other entrepreneurial investors. They established strong IP positions protected by the patent system that has so successfully ensured success for Big Pharma, through which they could establish licensing and other deals with the likes of Eli Lilly, Pfizer, the newly created Sanofi-Aventis, Merck, GSK, Roche, Monsanto and DuPont to name but a few.

Persidis (1996) described bioentrepreneurship as the wealth creation derived from the application of the biosciences to the business context. Bioentrepreneurs look for commercial value in every aspect of technology that they utilize. Innovativeness is vital to the creation of a biotechnology venture while credibility remains the backbone of the bioentrepreneur's character. The challenges of financing from venture capitalists prove to be a constant struggle for the bioentrepreneur. Similarly, risk-taking comes from dealing with the uncertainties of R&D, a rapidly evolving marketplace and the nebulous field of intellectual property.

Therefore, it appears at first glance, that traditional entrepreneurship has many similar characteristics to that of the new age bioentrepreneur. Yet, this deduction seems premature as there are other variables involved. For example, their backgrounds, education and organizational processes have yet to be taken into account.

PROFILING THE BIOENTREPRENEUR

In arriving at a profile of the bioentrepreneur, having just gained a feel for the industry in which they operate, we must first define the broader concept of the technopreneur. The technopreneur is a breed of entrepreneur that has emerged in the last two decades as a part of the global era of knowledge economies and new technologies. Technopreneurs recognize the value of technology and invent new ways of using it to change their industries (Teague 2000). They also make much use of computers, the Internet and other modern advancements in developing the business venture and its operations. Their skill sets are therefore predominantly technical rather than business oriented. Profiles of technopreneurs illustrate certain similar attributes to traditional entrepreneurs such as innovation, opportunity identification and capture, risk-taking and perseverance. Bioentrepreneurship is essentially a subset of technopreneurship (Figure 2.1), that is, technopreneurs that have begun their ventures in the biotechnology industry.

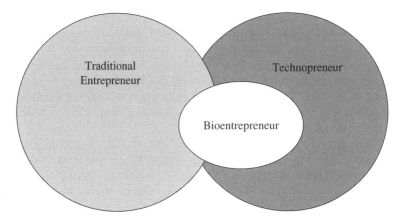

Figure 2.1   The overlap between the entrepreneurial forms

Bioentrepreneurs as individuals seem to sit at the highest end of the educational spectrum. The vast majority of biotechnology businesses are started up by holders of science degrees, MBAs, PhDs and various other tertiary qualifications (Janeitz 2003; Spence 1997). Their ages tend to be much older at the initiation stage of the enterprise as compared with the traditional entrepreneurs. This is mainly due to the length of time spent in university, as well as the time required to develop the idea into a viable product. It also appears that the majority of these entrepreneurs begin with education directed towards the sciences and therefore have an in-depth knowledge on their very complex products. Intellectual capital and subsequently property therefore commonly resides in the very heads of bioentrepreneurs, and can be the source of a biotech firm's competitive advantage (Schneider 2002). This phenomenon is not often described in traditional entrepreneurship.

Technical skills get the bioentrepreneur established in their venture. However, technical skills alone will not offer sustainable competitiveness, as exploiting opportunities relies heavily on traditional entrepreneurial skills. So the question remains. How does the bioentrepreneur or the entrepreneurial biotech company build this knowledge base of business strategic skills? Largely it has been through experiential learning rather than through the codified knowledge developed through university courses, although the trend toward the biotech PhD, augmenting their qualifications with an MBA is becoming apparent.

One of the problems with experiential learning is that experiences by their nature are limited. Most bioentrepreneurs commence this stage of their career late, having been successful in their scientific field first and after having completed the initial PhD. Their range of business experiences is often limited to one or two companies. This limited access to distributive knowledge can be problematic to building an entrepreneurial knowledge base not only in individual biotech companies, but for the industry as a whole. Such a dilemma is exacerbated by the barriers to potential entrepreneurs, particularly those from other industries, to entering the industry in a competitive manner, given the strong requirement for technical knowledge, as well as market knowledge.

Limited experiential learning experiences do not support robust strategic judgement, particularly in an industry where change can be considered a constant rather than a variable. The extent to which this has contributed to the high (and growing) failure rate of small biotech firms or whether this is an inevitable consequence of a long, complex and very expensive R&D pipeline is almost impossible to delineate. On the positive side, failure is a critical ingredient in the experiential learning process. The more failed entrepreneurs who come back to try again, the more robust the entrepreneurial knowledge base will be come in the industry.

With regard to the bioentrepreneur's environment, Genentech provides an interesting example. In the creation of a venture that explored the uses of gene splicing, Swanson had to convince renowned scientists to join this experimental company whilst trying to acquire venture capital in a new, remote area of business. Despite the reluctance of pharmaceutical companies to provide any financial support, Swanson went on to create a speculative market boom that inspired 'confidence' in the industry. Genentech's first major success was in licensing recombinant human insulin to Eli Lilly.

However the success of the bioentrepreneur and their company cannot be assured by the possession of the correct characteristics. The entrepreneurship must be combined with the desire and ability to innovate. Innovation and entrepreneurship go hand in glove. It remains unclear which is the hand and which is the glove.

REFERENCES

Andrews, P. (2002), 'The state of biotechnology in Australia', University of Queensland Lecture, Brisbane.

Barnard, R., C. Franco, P. Gray, D. Hine, W. Rifkin and F. Young (2003), 'A review of biotechnology skills in Australia', Canberra, Australian Universities Teaching Committee.

BIO (2001), 'Editors and reporters guide to biotechnology', A report by the Biotechnology Industry Organization, USA.

BioBusiness (2001), 'Landmark discoveries in biotechnology', Available: http://www.biobusiness.com, pp. 22–3, Accessed 16 March 2004.

Champion, D. (2001), 'Mastering the value chain', *Harvard Business Review*, June, 109–15.

Ernst & Young (2000), 'Convergence: The biotechnology industry report', Ernst & Young.

Ernst & Young et al. (2002), *Growing Our Knowledge Economy*, Melbourne: Ernst & Young.

Ernst & Young (2003), 'Beyond borders: The global biotechnology industry report', Ernst & Young.

Hine, D. and A. Griffiths (2004), 'The impact of market forces on the sustainability of the biotechnology industry', *International Journal of Entrepreneurship and Innovation Management*, **4** (2/3), 138–54.

Janeitz, E. (2003), 'Biotech + medicine', *Technology Review*, **106** (8), 72–9.

Mazzarol, T., Volery, T., Doss, N. and V. Tuein (1999) 'Factors influencing small business start-ups: A comparison with previous research', *International Journal of Entrepreneurial Behaviour and Research*, **5** (2), 48–60.

Persidis, A. (1996), 'Innovation: building molecular value', *Journal of Business Strategy*, **17** (2), 18–21.

Powell, W.W. (1998), 'Learning form collaboration: knowledge and networks in the biotechnology and pharmaceutical industries', *California Management Review*, **40** (3), 228–40.

Schneider, J. (2002), 'Intellectual property: the driving force for growth and funding', *Journal of Commercial Biotechnology*, **8** (4), 320–24.

Spence, R. (1997), 'The science of business', *Profit*, **16** (4), 53–8.

Teague, P.E. (2000), 'Technopreneurs', *Design News*, **55** (14), 9–10.

Trudinger, M. (2002), 'Jacobsen defends deal', *Australian Biotechnology News*, **38** (38), 1–5.

Woicheshyn, J. and D. Hartel (1996), 'Predicting value-added progress of biotechnology firms: An exploratory study', *Journal of Engineering and Technology Management*, **13**, 163–87.

3.   Innovation and R&D management

Co-authored by Nicky Milsom, Koala Research Ltd

INTRODUCTION

This chapter covers key aspects of the management of innovation and R&D in biotechnology, including acquiring technology processes, project planning and management, funding R&D projects, capital and the equity gap, venture capital, and managing innovation risk and uncertainty.

The management of the innovation process is vitally important, but provides a major challenge, with many elements to the process. Effective management of R&D and innovation are crucial to a biotechnology venture's success and survival, and are therefore a common management challenge in all biotechnology companies. Approaches vary according to size, aspect of R&D emphasized, and to some extent personal management style of managers in the company.

Many biotechnology ventures arise from organizations that place a significant emphasis on heroic/basic/blue-sky research, for example, universities, medical research institutes and other public sector research agencies. In the private sector, businesses that conduct research, such as pharmaceutical companies, also devote significant resources to applied and developmental work.

Many biotechnology firms may be less familiar with the demands of development. However, in order to be successful, the biotechnology venture needs to blend and manage both basic research and development appropriately, with a strategic view to commercialization, either independently or more likely as part of an innovation collaboration, such as a network or cluster.

An important aspect of the innovation management process for entrepreneurial biotechnology firms in a highly connected networked industry, is managing external collaborations that play a key role in the R&D process. Extensive interdependence adds to the complexity of managing the innovation process and developing an innovation strategy in a biotechnology firm. The founder bioentrepreneur plays a key role in this process.

ACQUIRING TECHNOLOGY PROCESSES

The management of the innovation process is a particular challenge in new biotechnology firms. In this context, innovation is defined as the process whereby a discovery is taken to the market for some commercial outcome.

The R&D process that relates to incremental innovation is necessarily a more systemic, management-oriented approach, which follows defined procedures in both the research and development phases. This is not to say that creativity is not still an essential feature of the process, it is simply that the combination of innovation and management tasks create a more structured approach to R&D. Blue-sky research cannot be restricted by such processes because the outcome is unknown. It is a more completely creative process. Definitions of the types of research undertaken by biotechnology firms are given in the next section.

There is some evidence from innovation research that a key difference in the innovation process is the driver for innovation within the industry. From this, incremental innovation is frequently driven by users whereas industries experiencing radical innovation are largely driven by the R&D (Brusoni and Geuna 2004).

R&D underpins the biotechnology industry; it is one of the most research-intensive industries in the world. According to the Biotechnology Industry Organization, Expenditure on R&D by US biotech firms increased from US$5.7 billion in 1993 to US$20.5 billion on R&D in 2003, with the five biggest firms spending on average over $100 000 per employee on R&D (BIO 2004).

Most biotech firms are therefore highly research intensive, but also under pressure to innovate, that is produce commercial products, processes or services from R&D within the timeframe of the funding horizon. Managing the innovation process (including obtaining timely, appropriate funding for all stages of the discovery pipeline and generating commercial outcomes) is crucial to ensure firm survival.

Managing the innovation process therefore requires not only excellent R&D but also the development and release of final products or processes or services. This is very different to the normal pressures experienced by the bioentrepreneur, for three main reasons. First, the bioentrepreneur is frequently from either a research institution or a spin-off from a public sector research organization. Ernst and Young (2003) estimated that over 50 per cent of Australian biotech firms are generated from public sector research organizations. While Druilhe Garnsey (2004), estimated that a significant proportion of new firms in Cambridge, UK arise from 'serial' entrepreneurs who are familiar with the challenges associated with acquiring ongoing

R&D funding from a variety of sources. The industry in some countries is too young to have serial entrepreneurs who have had success with previous ventures. For example in Singapore the number of successful bioentrepreneurs is limited by the very small number of biotechnology companies which have been in existence for more than five years. The extensive history of industry generally and biotechnology specifically around Cambridge creates the environment where serial entrepreneurs, and biotechnology business angels emerge in the local environment.

Publicly funded research is still an important aspect of the funding regime for NBFs. The R&D grant process is to a certain extent predictable in that it operates to a regular calendar, with regular timeframes for funding openings and decisions. Second, the competitive R&D grants process is determined by the track record of the researcher and the quality of the proposal (generally as determined by peers); important outcomes include publications rather than commercial products or processes. Third, a grant is generally just that – it is funds that are handed over for a specific project or programme, and unlikely to be terminated mid-project unless there are exceptional circumstances, for example failure to report back.

Frequently the expertise of the new biotechnology firm is firmly grounded in the research side, with development and commercial skills missing. However, having an excellent basis in R&D is a definite advantage to the new firm – particularly if the founder and team are recognized internationally for their research excellence. This recognition will help the new firm attract partnerships with major companies, which in turn helps the new firm to acquire missing expertise through strategic alliance, merger or acquisition.

This may help overcome the biggest impediment to a biotech firm that is grounded in the culture of a research institution: the commercial imperative. Firms must be able to remain clearly focused on track to produce a product, process or service, be it a drug, device or new plant variety. This single focus becomes more important once investment capital is secured and commercial milestones agreed. One of the key challenges in research and development management for the bioentrepreneur is therefore to encourage the dual focus of (blue-sky) research whilst simultaneously achieving commercial outcomes by the research team.

This challenge is more difficult if the young company retains strong links with a host research institution, and more particularly if the research staff have dual academic and commercial roles, but is exacerbated further if the founder and bioentrepreneur also has a dual role.

In the next section we explore the different types of R&D and where the biotech firm fits in this process.

DEFINING R&D PROCESS TYPE

R&D applies to the process of development from idea generation to the point at which a tangible product or a service is ready to be brought onto the market. The OECD defines Research and Development as 'creative work undertaken on a systematic basis in order to increase the stock of knowledge, including knowledge of man, culture and society, and the use of this stock of knowledge to devise new applications' (2002, p. 30).

In addition to this definition, the OECD discriminates between basic and applied research and experimental development.

- *Basic research* is experimental or theoretical work undertaken primarily to acquire new knowledge of the underlying foundation of phenomena and observable facts, without any particular application or use in view.
- *Applied research* is also original investigation undertaken in order to acquire new knowledge. It is, however, directed primarily towards a specific practical aim or objective.
- *Experimental development* is systematic work, drawing on existing knowledge gained from research and/or practical experience, which is directed to producing new materials, products or devices, to installing new processes, systems and services, or to improving substantially those already produced or installed. R&D covers both formal R&D in R&D units and informal or occasional R&D in other units (OECD 2002, p. 31).

There is a very distinct difference between heroic/basic/blue-sky research that is largely undertaken by public organizations and incremental/applied research and development, which is the main focus of most industry.

In the biotechnology industry, firms have to conduct a range of R&D – from blue-sky through to applied research and experimental development. This raises a major challenge to the research team, particularly one unfamiliar with the demands of strict milestones associated with applied and experimental development. Such a focus requires excellent project management skills that ensure attention to timelines and adequate planning while simultaneously encouraging blue-sky experimentation.

Before exploring types of R&D and equity funding available to the biotech firm, it is important to understand the model of innovation typically pursued in the biotechnology industry.

Models of Innovation in the Biotechnology Industry

We have previously noted the highly research intensive nature of the biotech industry. R&D provides a major source of a firm's overall expenditure, and consequently leads to the extremely high 'burn rates' of new firms.

The biotech drug discovery process is typically described as a modified linear process, as indicated in Figure 3.1. This is somewhat typical of the life sciences industries generally that are driven by findings from R&D. Human health applications of biotechnology, in particular, are characterized by very long lead times to production due to extensive regulatory requirements (see Chapter 9). Applications in agribiotechnology are also subject to regulation and testing before approval for release onto the market, whereas environmental biotechnology is constrained to a lesser degree by need for testing and approval.

There is some debate over whether post-market surveillance is actually part of the R&D process, although it does impinge on the final success of the innovation process.

Generally, the shorter the approval time, the less the return; high-risk products, such as new drugs, if successful, generate a much higher return than a medical diagnostic test, which has a much shorter lead time.

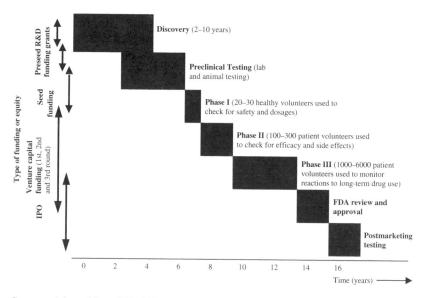

Source: Adapted from BIO, 2004: EVCA, 2004.

Figure 3.1   Drug discovery process and type of funding

Approval times for new drugs in the USA have been increasing, from just over nine years in the 1960s to over 14 years in the 1990s (Burrill 2004). This is despite the 'quicker' drug testing and approval process introduced by the FDA and described in more detail in Chapter 9.

It should be noted that the timeframe for the approval process may vary between different countries, with Japan and European approval times for new molecular entities taking longer than the USA (Japan by a factor of almost two, and Europe by 1.5).

Figure 3.1 also shows typical sources of finance for the biotech firm as it moves through the various testing stages. In the next section, we explore these in more depth. We begin by considering R&D funding, and then various types of seed and equity finance offered by private and public sources. Finally we consider the firm becoming publicly listed on a stock exchange through an initial public offering (IPO).

FUNDING THE BIOTECHNOLOGY COMPANY: TYPES OF RESEARCH FUNDING

Research Grants

Research grants for basic research are generally funded by public and public not-for-profit sources. This is generally because the research is considered as too high-risk and long-term for industry to sponsor. Added to this, the serendipitous nature of biotechnology research means the product outcomes, and the eventual target industry sub-sector, are too unpredictable for private investors to delve in too early. Governments in most countries are the primary providers of research grants through block funding awarded to research institutions, as in the UK and Australia, and/or competitive research grants 'won' on the basis of research excellence (for example, NSF and NIH (USA); Research Councils (UK); ARC (Australia); MORST (New Zealand); and Research (Canada). Biotechnology makes up a growing proportion of public R&D funding, being an estimated 13.8 per cent in Belgium, 10.1 per cent in Canada, 8.1 per cent in Finland and 7.8 per cent in the UK and Australia (OECD 2001).

Research grants are frequently referred to as 'sponsored research' in the USA and Canada. A sponsored research agreement is defined as 'an agreement between the Sponsor (agency) and the Researcher (PI) that has a clear statement of work and set of deliverables in exchange for which the PI receives an agreed upon level of support for a specified period of time' (CALTECH 2004).

Table 3.1　Characteristics of sponsored research

- Peer review
- Assessment based on:
  - Excellence
  - Track record of researchers
  - Significance and innovation of project
  - Contributions of participants (for applied/partnership applications)
- Provided for disciplinary applications
- Mission-oriented research agencies match the aims of research to those of the specific agency (e.g. US Department of Agriculture)

Sponsored research grants have unique characteristics (Table 3.1) that separate them from industry funding.

The main risks associated with research grants are related to juggling income from a variety of sources; ensuring a sufficient pipeline of applications and grants; making up funding where there is a shortfall as most funding agencies do not award the full costs of research; publishing findings on a regular basis and effectively managing competing demands for time including teaching, supervision, administration and research for university staff.

Public institutions frequently are the main source of biotechnology entrepreneurs. This gives rise to a significant tension for the future development and growth of the company, since the drivers for academic success are very different to those in industry. Key to academic success includes ability to publish in highly reputable journals, and attract competitive grants.

This grant mentality – that research is funded based on peer reviewed applications that provide funds to cover costs of the research project itself (based on applications for projected research rather than milestones), and assuming that the host institution provides the background (or indirect) infrastructure and support is a major issue for bioentrepreneurs from public research institutions.

The tension is especially strong for those researchers that are still in the process of establishing an academic career, where 'traditional' track record is crucial to future internal and external success (including tenure). Tenure itself provides a further complication by appearing to offer a 'safe' place in an organization that is not replicated in the start-up. Frequently a good researcher in a public research organization, particularly a university, is one that can develop a systematic mechanism to identify grants to successfully apply for, which will cover the various stages of a project or programme of research.

Research grants also involve little personal or reputational risk – other than the ability to satisfy the granting body that the work has been conducted according to the project schedule, and appropriate reporting mechanisms followed.

The process of moving from the research programme/project once the IP is established and protected is a complex one for many new biotechnology firms. Essentially the NBF comes into existence once the decision to commercialize the IP is taken. Some choose to go it alone, undertaking substantial risk – particularly financial risk to the founder(s). Many however will find it very difficult to face this kind of risk and choose not to stray far from their origins. In doing so they keep their venture more simple in design and operation, however they also find it far more difficult to differentiate between a research project and a business.

Increasingly however, the best researchers also excel at recognizing the commercial potential of their research, and protecting and exploiting the outcomes. However, if such 'star' performers (Darby and Zucker 2002) are to become bioentrepreneurs they must usually make a conscious career choice – leave academe (and its relative security) for the roller coaster ride that is a start company. In some cases a host institution may make the choice easier by providing periods of secondment or a percentage of time recognized as being free from normal duties while the new firm is being established. In other cases, the researcher may continue to do both; this may be a successful scenario if the full support of a high quality, well resourced technology transfer office is given in establishing the company.

At some point the choice between a full-time research career or one in a start-up company must be made. This may be softened by the researcher moving to head the R&D activities of the new firm, allowing continuation of the research career, while an experienced CEO is brought in to manage the company (perhaps on a 'sweat–equity' basis).

In other cases, the break with the research institution may be total. Under such a scenario, the new firm must succeed or die, as there is no institution to prop up problems or lack of cash flow. The major decision points to be made are outlined in Figure 3.2.

Where accessible, sponsored research provides funds to the new biotechnology firm to allow conduct of costly R&D. However, biotechnology firms are often not eligible alone to seek sponsored research funding, therefore they must partner with a university or public research organization. In addition, sponsored research grants do not usually cover the full cost of research where the government is trying to obtain 'a circulation of ideas, capital and talent' (Hamel 1999).

Once the fledgling biotechnology company has moved from the host institution to a separate company there are constraints about ability to

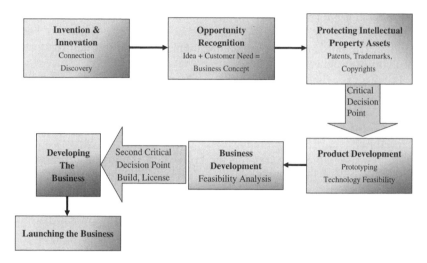

Figure 3.2   Decision points in the commercialization process

access public grants for R&D without a partnership in place. R&D grants schemes that do operate frequently require the firm to provide some form of matching cash. One exception is the US-based Small Business Innovation Research (SBIR) and Small Business Technology Transfer (STTR) programmes, which are the world's largest seed capital fund for small technology companies, with $525 million a year available for biomedical and behavioural research.

One of the advantages of a biotech team having a strong, demonstrated basic research record is that it highlights the firm's ability to create new knowledge and be at the forefront of its field. This can act as a major draw card to industry, including Big Pharma, who are increasingly contracting out high-risk, early-stage research to start-ups and research organizations, as discussed in the next section. As we shall see in the following sections, a strong track record in research may also help convince investors of the potential of the idea, although this is highly unlikely to be the sole criterion used for investment-type decisions.

One of the major issues facing NBFs is the increasing complexity they face the further they advance from their host institution or organization. Considerations in a research grant are relatively limited, offering a simplified perspective on the world for those leading a new firm with little beyond their scientific skills base to support them. By contrast considerations in developing and implementing a business plan are vastly more complex. The business planning process, while often tackled too late and

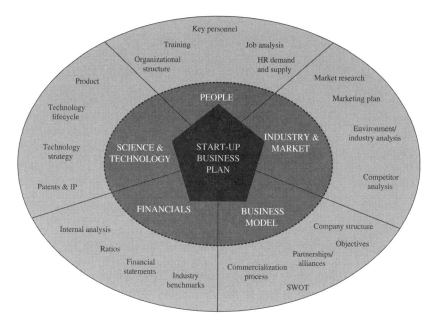

Acknowledgement: Thanks to Paul Curwell, Office of the Queensland Chief Scientist for the development of this map.

Figure 3.3   Considerations in the biotechnology business plan

drafted by consultants rather than those inside the firm, is a fundamental step in the early development of an NBF. The abundant new terminology and decision making criteria are daunting even to the experienced bioentrepreneur. A view of Figure 3.3 offers a glimpse of this complexity.

Looking at the value chain of a biotechnology company in most countries with a burgeoning biotechnology industry there is a source of government funding for almost every stage of the value chain. This means that a NBF could potentially gain substantial funding sufficient to take it well on the way to growth. It may in fact avoid 'the market' completely for the first few years of its existence. This mitigates financial risk especially to the founders – this is something which sets them apart from most new venture founders – the financial risk is often not theirs.

Funding for Applied and Contract R&D

NBFs, because of their obvious commercial potential have been provided substantial governmental and institutional support mechanisms not afforded to most other industries. The result: a myriad of programmes, funds, schemes and bodies ready willing and able to throw money at the NBF. However, one of the most important factors in the continued growth of the biotechnology industry is the increased availability of pre-competitive/contract research funds from pharmaceutical companies.

Big Pharma in particular, has moved to outsource high-risk, early-stage research to small biotech firms. This 'innovation gap' reflects Big Pharma's drive to reduce the difference between R&D spending and fewer drug approvals (Burrill and Haiduck 2004). However, there is evidence that deals between pharmaceutical companies are becoming fewer but larger and more complex (Haiduck and Burrill 2004). The ability to negotiate and close a complex deal may be a substantial challenge to a new biotech firm without commercial/legal skills, and drain scarce resources if costly external legal advice is sought.

A major criterion in a biotech firm being able to trigger such funds is the quality and excellence of the research team, as previously discussed. This may be a disadvantage in pushing a research-institute based team back to a grant mentality (but where funds are subject to strict commercial-type milestones) rather than assisting in the development of a business culture within the firm. Careful management of such a process is therefore important, and here advanced project management skills come into play.

PROJECT PLANNING AND MANAGEMENT

Most biotechnology innovation is a story of product development, innovation and diversification, all aimed at improving return on investment and even profitability. New product development (NPD) can be treated as a project and therefore requires project management skills, but many of the start-up companies are considered as projects and hence the management of the entire companies requires well-honed project management skills.

Project management (PM) has been defined as:

The application of knowledge, skills, tools, and techniques to project activities in order to meet or exceed stakeholder needs and expectations from a project. Meeting or exceeding stakeholder needs and expectations invariably involves balancing competing demands among:

- Scope, time, cost and quality.
- Stakeholders with differing needs and expectations.
- Identified requirements (needs) and unidentified requirements (expectations). (Project Management Institute, Standards Committee 1996, p. 3)

The project management triangle consists of cost, time and performance. The implication of this structure is that the three are inextricably linked. If costs are reduced, either time or performance suffers. In order to accelerate the project, more resources and a concomitant increase in cost will be required. An increase in the performance (scope) and cost or time or both must change.

The Importance of Planning in the Project Management Process

Planning is a cornerstone of project management. It is a vital part of the project management process and for this reason alone it is one of the most important functions. Planning activities can vary considerably, depending on the type of project and the management environment in which the project originates and develops. They can range from a simple descriptive coverage of what needs to be done to extensive computer-generated diagrams and printouts showing details of every aspect of the project. Poor planning has often been blamed for poor project performance and management problems. Planning needs commitment and ownership, and an appropriate attention to the level of details required. Where planning is linked to future funding, proposals such as business plans need to be made as explicit as possible by incorporating factors such as markets, finance and so on.

Planning is frequently neglected or inadequate in the new biotech company. However, there are two elements to the input side of project planning that can be drawn in to the earliest stage company:

- Experienced staff who can draw on their own direct knowledge of successful and unsuccessful outcomes in past projects; and
- The appropriate tools and techniques by which this experience and creativity/innovation can be converted to plans, data and motivation for those associated with the endeavour.

Participation

Planning cannot be done successfully in an ivory tower or in a vacuum. It is the overall collective responsibility of the project manager and his team of specialists – whether they are a dedicated team or a mix of staff from functional areas. One person does not have all the specialist knowledge to adequately plan all areas of a project. Each specialist has his or her own perspective of the project and this can often be a narrow perception related to their own area of expertise. This perception has to be drawn out and developed into its correct place in the overall plan.

Members of other agencies and major contractors, where appropriate, who will/could have a major involvement in the project, will need to be consulted to provide at least basic data on time-scheduling of their contribution. This is likely to reveal interesting dependency linkages that may not have been evident initially and it may introduce some new 'players', who were not envisaged.

Level of Detail Required

The level of detail required depends on how much control or accuracy is needed by the project manager, how much visibility or monitoring is required and how much risk is expected to be reduced if the additional analysis is achieved and the commensurate detail included in the planning. One of the undisputed benefits of doing explicit detailed planning using techniques such as the critical path method (CPM) is that it enforces a discipline in which all aspects of the project are defined, analysed and connected – providing an assurance that important aspects, and even seemingly unimportant elements, will not be forgotten.

How Planning can be Achieved

Overall, planning should be uniform in approach and structure – where a bureaucratic organization deals with many projects at the one time, this uniformity is critical for the efficient handling of the total programme.

THE BIOTECH FIRM AND THE R&D CYCLE

It is important to understand where the biotech company stands with regards the R&D cycle and how much of the cycle it attempts to cover. The smaller the biotech company, the fewer elements of the R&D cycle it will have the resources in which to compete. Outside the USA, there are few fully integrated biotechnology companies, that is firms that are able to cover all aspects of the process from discovery through to manufacture and marketing.

R&D cycles will be explored in more depth in Chapter 4. However two things must be understood by anyone managing the innovation process in biotech companies:

- The length of the R&D cycle for each product being developed by the company;
- The elements of the R&D cycle for which the company is responsible.

It is worth noting that 'excessive use of external sources can also reduce the investments in developing internal manufacturing capabilities, which can weaken a firm's competitive position' (Zahra and Nielsen 2002, p. 380).

The proof of concept (POC) stage of development includes all the activities required to establish the commercial value of the proposal. These activities will vary with each project, but they will typically aim to establish the validity of the discovery, to prove its efficacy or to demonstrate the commercial applicability of the project.

As the majority of innovation occurring in industry is incremental, it is this area of R&D management that will be concentrated upon. This also allows other existing management techniques to be applied in the R&D realm such as project management, as the development of each new product has a distinctive timeframe that is the basis of projects.

THE RELATIONSHIP BETWEEN THE RESEARCH BASE AND INDUSTRY

In smaller economies such as Australia, New Zealand and Singapore, the relationship between public institutions and the research base of the country is extensive. There is also less involvement of the private sector in discovery and basic research, so this is undertaken by universities, hospitals and publicly-funded institutes. As a result the commercial opportunities that emerge, while needing to be developed in a market environment, owe much to their institutional roots. It is often difficult for the biotech venture that has spun-off from university research to change its culture and strategy toward a fully commercial focus.

In some cases, the host research institution continues to provide an umbrella for the new firm, and the spin-off may also come to rely on the institution for basic administrative support. The host institution not wishing to let the spin-off fail exacerbates this.

The scientist, and more particularly star scientists, play a vital role in knowledge transfer to the firm, particularly when they create ties to the firm in terms of equity or exclusivity agreements (Zucker et al. 2002). In addition, involvement of the star scientist in the development and growth of the firm is a key component to firm success.

The lure of further public funding, and the comfort zone for scientists of applying for grants to support their research makes the break between the biotech venture and the public institution difficult to achieve. This is particularly the case when the host public sector research institution supplies most of the resources such as, labs and equipment, technicians, undergraduate and post-graduate students. It is akin to the new venture raised in an incubator, research or technology park having to graduate. The experience in incubators is that many of the graduates had to be forced out. It is a matter of being cruel to be kind. There are limitations and restrictions inherent in public institutions that limit the commercial viability of new biotech ventures.

So when should the fledgling leave the nest, how much association should it retain and how does it build the strategic alliances that will broaden its outlook and commercial opportunities? Proteome Systems (PSL) from Australia, a previously unlisted public company, which listed in 2005, is a good case in point. They were going to be denied contract research funding by their host university until they ceded from the university. PSL were probably lucky in this regard, in that the opportunity was recognized externally and support offered. In many cases however, opportunity recognition is serendipitous. It can happen at conferences, dinners, seminars, chance meetings in airport lounges or hotels, newspaper articles and other haphazard means. Types of equity and investment funding available to the biotech firms are discussed in the next section.

CAPITAL AND THE EQUITY GAP

Defining Capital

As a first step in looking at venture capital it is important to look at the three major types of capital:

- *Fixed capital*   This is needed to purchase the business's permanent or fixed assets. These assets are used in the production of goods and services and are not for sale.
- *Working capital*   This represents the business's temporary funds; it is the capital used to support the firm's normal short-term operations. Working capital is normally used to buy inventory, pay bills, finance credit sales, pay wages and salaries and take care of unexpected emergencies.
- *Growth capital*   These requirements surface when an existing business is expanding or changing its primary direction and seeking new markets.

However, when considering funding for high technology ventures, stage of funding, and therefore size of monies and stage of development, more frequently becomes the defining criteria.

The Equity Gap – the Reason for the Venture Capital Market

Venture capital arose primarily as a result of the inability of smaller high-tech firms to source funds from more traditional sources. Lending to small business has been unprofitable for many banks due to low margins and high losses in the past. Also, banks use traditional measures in their lending decisions that often preclude the newer, smaller, high-risk firm.

A traditional approach to the lending decision by banks is to require security in the form of a house, vehicles and other associated assets that are not directly associated with the business, in the case of sole traders and other unlimited firms. Banks also want tangible assets in the form of equipment buildings and land. They often ignore the valuable intangible assets these firms have in the form of IP, expertise, goodwill and reputation.

On this basis, banks will not lend to many new ventures and therefore a gap is created between what is needed to establish or grow the firm and what is available. This is the basis for the establishment of other forms of equity, including pre-seed, seed and the venture capital market to fill this gap.

A notable exception to the traditional bank approach occurred in the Cambridge region, where one bank manager decided to provide small portions of early seed funding for new high-technology firms spinning out of Cambridge University (Druilhe and Garnsey 2004).

Globally 44 per cent of funding deals are done at discovery phase; 16 per cent at pre-clinical; 10 per cent at phase I and II; 6 per cent at phase III; 5 per cent at registration and 19 per cent after launch (Allant CBIM 2004). Figure 3.4 highlights the typical equity and investment funding through the growth stages of a biotech firm.

The funding can essentially be broken down into several stages, each with distinct characteristics.

a. Informal funding: founders, family, friends and fools
b. Pre-seed and seed funding: business angels, pre-seed funds, for example, funds provided by government or public–private partnerships
c. Venture capital funding
d. Private equity investors, mainly for expansion or management buy-outs
e. IPO

Each of these sources has different criteria, motives and levels of involvement in the biotech firm. Funding profiles also differ between different regions, and between industries of different maturity. Some of these issues, according to funding source, will be discussed below.

Most important, it should be recognized that not all biotech firms can hope, or necessarily want to, trigger the full variety of funding sources

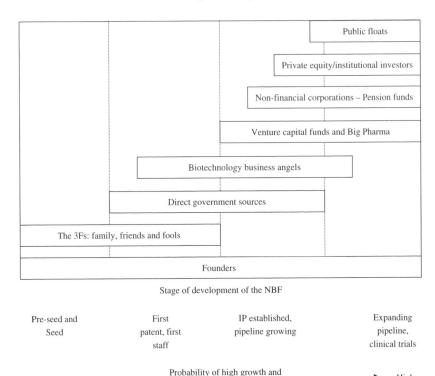

Stage of development of the NBF

Pre-seed and	First	IP established,	Expanding
Seed	patent, first	pipeline growing	pipeline,
	staff		clinical trials

Low ⟵————   Probability of high growth and   ————⟶ High
                           returns

Figure 3.4   Sources of equity and investment funding for biotech firms

available. Ideally the push should be to generate revenue from creating new products, services or processes that can generate sufficient returns to allow the firm to grow. The mix of funding for the biotechnology industry in the USA is given in Figure 3.5.

THE STAGES OF BIOTECH FUNDING

Founders, Family and Friends

Primary funding sources of a new venture include the founder, family and friends. In the case of a new biotech venture originating from a host public research organization, the institution itself frequently provides some early stage funding, rather than the founder, family or friends. This may be in the form of cash, in-kind such as use of facilities at little/no cost, or

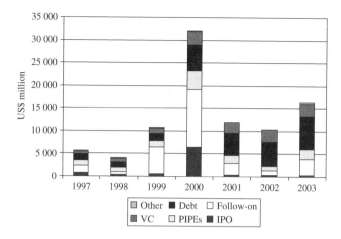

Source: Haiduck and Burrill (2004).

Figure 3.5 Funding sources for the US biotech industry

commercialization/patenting advice, or a mix of both. In most cases this early funding is provided in exchange for some equity in the firm, and possibly a position on the board of directors.

One of the issues with the new biotech firm is its high burn rate due to the need to conduct expensive R&D, and the need for equipment and scientific expertise. This frequently leads to a significant gap in funding between the amounts that can be raised by 'traditional' funding sources of family, friends and founders and the amount needed.

Pre-seed and Seed Funding

A number of governments have recognized this gap, and make available grants to assist the new company establish a firm commercial base, for example by providing grants to assist with business planning and marketing expertise.

A key ingredient to success is the management and commercial ability of the research team. For a new biotech firm founded by a researcher, it is difficult to attract sufficient funding to cover the costs of R&D, let alone hire expensive management teams. In some cases, the host research institution's technology transfer office may provide some assistance and help put the company in touch with an experienced business angel or manager who is prepared to put in time with the firm in exchange for equity – a 'sweat–equity' CEO or manager.

One of the issues associated with venture capital investors funding early stage firms is that the criteria used to make funding decisions frequently requires in-depth due diligence which simply may be inappropriate for an early-stage biotech firm. Such a mismatch may lead to a general surplus of funds available, yet with early-stage ventures finding it difficult to acquire funding. This situation is exacerbated in less mature biotech industries where there is also a lack of critical mass of experienced analysts and investors, such as Australia, New Zealand and parts of Europe. In such cases, having a strong pool of informal investors, such as business angels, becomes even more important.

Business Angels

A business angel is an individual of high net worth, who has often generated their wealth through creating their own successful start-up venture(s) (BVCA 2003). Where such individuals choose then to invest in other new companies, they will frequently choose firms from the same sector.

The maturity of a region's biotech industry frequently determines the availability of business angels. In a mature industry such as the USA, where there are over 1400 firms (BIO 2004), there are a much larger proportion of angels, particularly in the major biotech hubs such as San Francisco, Boston and North Carolina (Ernst & Young 2004). However, in maturing and immature industries there is a much smaller pool of angels.

It is regarded that the informal venture market is significantly larger than the formal market and that business angels fill the so-called equity gap by making investments precisely in those areas in which institutional venture capital providers are reluctant to invest. However precise measures of amounts of angel funding are difficult to ascertain due to the informal nature of these investments. It is estimated that the typical business angel in the UK makes one or two investments in a three-year period, either individually, or through linking up with others to form an investment syndicate (BVCA 2003).

The lack of business angels can also have significant implications for management expertise available to new firms. Some governments have actively sought to encourage business angel involvement in high-tech start-ups, including biotechnology. The Scottish Enterprise Council provides such an example, where government funding of up to £50 000 is matched for approved consortia of business angels and early-stage investors (Scottish Enterprise Council 2004). Such schemes are not only intended to raise availability of funds for new ventures, but also to increase availability of management expertise in new firms.

Government involvement in early-stage funding must address account-ability issues associated with investing in private for-profit ventures, and needs an appropriate framework in which to operate. One of the drivers of co-investing in such schemes is the leverage that the government may achieve on its funds. Ideally, substantial leverage should be obtained (by a factor of two or three on the public investment). In addition, business angel criteria for funding may not be as precise and definable as those used by more formal investment houses.

Venture Capital Funding

Venture capital funding is generally in the form of equity. Venture capital-ists manage funds, generally on a hands-off basis on behalf of other investors with the specific aim of making high returns. The other investors they might act for may include banks or superannuation/pension funds. Most venture capital funds expect to return a yield by exiting from their investment within five to ten years.

Many new high-tech firms, particularly those reliant more on a service than a manufacturing base, are not built on a substantial asset base. This is particularly true of the new biotech firm, although in some cases investors are prepared to consider patents as assets in the more traditional sense.

There are three major categories of businesses that seek venture capital. These are seed, start-up and growth firms. Start-up firms are probably the most common in biotech. They are firms in which the owner has a new, usually innovative, product that requires financial support for its commer-cialization. These are often considered to be high-risk ventures because they haven't proven themselves in the market.

Venture capital investment decisions take considerable time (from two to 12 months), and may involve lengthy due diligence processes and negotia-tions regarding equity and milestones. In many cases the venture capitalist will seek to play a significant role in the operation of the firm, such as a seat on the board and possibly day-to-day operational management involve-ment too. The degree of intervention of the venture capitalist in the running of the business is dependent upon the degree of development of the business concept. This is explained further in the next chapter.

Frequently, venture capital managers will seek to syndicate their invest-ments by bringing in other investors to spread the risk and add to their pool of expertise. The fund managers will also generally take a portfolio approach to investment by mixing their investments across a range of firm types and stages, again to reduce risk. Generally the degree of expertise required results in focus on a specific industry sector such as drug develop-ment, devices, or agricultural biotechnology.

Private Investors – Mainly for Expansion or Management Buy-outs

Other private investors consist largely of banks, financial institutions and other institutions providing debt financing both short and long term to businesses. Most private investments are provided for much later-stage growth or expansion, or to management buy-outs. These are much lower in risk as investments, compared to the high-risk, high return of earlier stage venture capital funds.

IPO

Initial public offering is where the firm lists publicly on a stock exchange. The biotechnology industry has had highs and lows in recent years with respect to the success of public floats; 1999 and 2000 saw some spectacular public offerings, with similarly major failures in 2001 and 2002. The market changed once again in late 2003 and 2004, with increased investor confidence and successful offerings.

It is worth noting that significant differences can be noticed between the maturity of different industries in IPO behaviour. In the USA, a mature industry, the average IPO in 2002 raised around US$70 million, around double that in the UK and ten times that in Australia (Burrill and Haiduck 2004). US firms are expected to be in final stages of clinical trials before listing; in the UK, at least to Stage III clinical trials, whereas Australian firms are more likely to be in pre-clinical stages (Vitale 2004).

Moving to IPO has significant financial and management demands for the firm. Public listing requires onerous reporting and accounting processes, as well as the firm having to satisfy shareholders' demands. Cost of IPO can also be significant.

The bioentrepreneur therefore has a number of choices on how to finance the various stages of the firm's growth. However, each type of financing has advantages and disadvantages that may have major implications for the future success and growth profile of the firm.

PREFERENCES BETWEEN DEBT AND EQUITY FINANCING

Debt Financing

This requires the venture to pay back loans with interest rates often higher than that charged to larger firms due to the risk factor in the lending. Debt financing does however allow the owner/management of the venture to maintain control of the business.

Equity Financing

This allows the business to grow without a significant debt burden to slow its progress and limit its cash flow. However equity capital usually comes at a price. Venture capitalists often require that they have significant voting rights in the company and may further insist that they take over the running of the company while the original owner is relegated to a technical position. It must be remembered that profit is the motive for most venture capitalists – they are providing capital to firms that could not obtain funding from other sources, therefore they need to control their risk by ensuring that the management aspects of the business are in their control. They are effectively then controlling the internal environment of the organization so that they can better cope with the turbulent external environment faced by many of the ventures.

MANAGING INNOVATION IN SMALL BIOTECH COMPANIES

In small firms with fewer resources, informal innovation may frequently be an important factor in building a successful company in the long term. The way that innovation is managed in a small firm therefore differs from that of a larger firm, but can be just as effective.

One of the founding companies of biotechnology, Genentech, illustrates how the integrated biotechnology company stresses the importance of performance across all functions, and not just R&D:

> Genentech, the founder of the biotechnology industry, is a company with a quarter-century track record of delivering on the promise of biotechnology. Today, Genentech is among the world's leading biotech companies, with 12 protein-based products on the market for serious or life-threatening medical conditions and over 30 projects in the pipeline. With its strength in all areas of the drug development process – from research and development to manufacturing and commercialization – Genentech continues to transform the possibilities of biotechnology into improved realities for patients.   (Genentech 2004)

Common challenges are experienced by small biotechnology companies, leading to management solutions and practices that are truly global. In reality these solutions may differ more according to the size of the company than between countries.

The case above indicates that the innovation management approach of biotech companies needs to match the company's core competencies – the greatest strengths of the company which are unique enough to provide a competitive advantage. The R&D function is like other functions in a business such as marketing, production, finance and human resources. As many firms see the opportunity to outsource the functional areas they do not perceive strengths, R&D could be one of these. Outsourcing this expertise reduces the cost of one of the most resource-dependent areas of running a business. So from a cost reduction perspective it makes sense to outsource R&D if it is not achieving desired results. Alternately it is difficult to achieve the same levels of commitment to products and to the company from another organization. However, the option to outsource is more relevant to integrated, larger companies; small biotechnology firms are more likely to make up missing expertise through alliances or partnerships (Burrill 2004).

Beyond the outsourcing of R&D, technology can be acquired from external sources that require integration into the firm. This is particularly relevant for the newly established biotech firm or one seeking to grow. These are often generic off-the-shelf technologies that are easily transferable between business environments. This is not always the case as technology that is quite sophisticated and difficult to implement can be acquired. While the cost of developing the technology in-house may be prohibitive, the cost of implementation of systems such as HR information systems often prove far more expensive to implement than planned. In some instances the technology is so incompatible with current systems that it is eventually abandoned as it fails to add value.

Decisions on outsourcing need to be taken very carefully in the context of appropriate project planning and management. As we saw in an earlier section, project planning is relevant for the new biotechnology firm as often the new firm's early stages are more like a project.

Innovation Risk and Uncertainty

One of the major challenges associated with the management of R&D and innovation is the management of risk. Uncertainty is an inherent characteristic of both. Management of uncertainty, and of decision making stop-gap decisions – for example relating to continuing, changing direction or particularly terminating a project are all major challenges to established R&D based industries, let alone ones which are characterized by rapid change and high cost R&D with lengthy time-frames.

Risk management planning

There are several risk management processes aimed at reducing risk. Turner (1993) provides a five-step process by which project risks can be managed:

1. Identify the potential sources of risk on the project.
2. Determine their individual impact, and select those with a significant impact for further analysis.
3. Assess the overall impact of the significant risks.
4. Determine how the likelihood or impact of the risk can be reduced.
5. Develop and implement a plan for controlling the risks and achieving the reduction.

On the other hand, Kerzner (1998) has a four-step methodology – risk assessment (which includes identifying and classifying risks), risk analysis, risk handling and lessons learned. The Australian and New Zealand standard (AS/NZS 1995) has a five-step process, including a preliminary step called 'Establish the context'. Under this heading, the standard suggests that the strategic, organizational and risk management contexts should be considered. The strategic context can include financial, operational, competitive, political and social aspects, to name but a few. It also includes the identification of stakeholders. The organizational context includes the organization's goals, objectives and capabilities and hence indicates how the project, and the project's risk, will be treated by the organization. In the risk management context, the scope and application of the risk management process are considered. The other steps in the standard are identification, analysis, assessment and treatment.

Considering the management of risk takes us back full circle to funding the biotech venture, with the entire process being one of identification of opportunity, analysis of options, assessment and strategies to achieve optimum funding profiles for minimum risk but maximum return for the firm.

CONCLUSION

We have covered plenty of ground in our look at innovation and R&D management. Raising funding for R&D, which is expensive and long-term, is a common challenge for all biotech companies, as is managing the R&D process to optimize commercial and research outcomes simultaneously. Approaches will vary according to the location of the industry, size of firm, aspect of R&D emphasized and to some extent personal management style of managers in the company, as well as the stage that the R&D has reached.

REFERENCES

Allant CBIM (2004), 'Global biotechnology industry deal structures', Working Paper, Lyon, CBIM.

BIO website (2004), Available: http://www.bio.org/speeches/pubs/er/statistics.asp, Accessed: 24 June 2004.

BVCA (British Venture Capital Association) (2003), *A Guide to Private Equity*, London: BVCA.

Brusoni, S. and A. Geuna (2004), 'An international comparison of sectoral knowledge bases: persistence and integration in the pharmaceutical industry', *Research Policy*, **32** (10), 1897–906.

Burrill, G. (2004), Biotech 90: Into the Next Decade.

Burrill, S. and G. Haiduck (2004), 'Life sciences . . . circa 2004: biotech's back on track', conference presentation, Kellogg Biotech Conference, USA, 17 April.

CALTECH (2004), Sponsored Research Office, Available: http://www.atc.caltech. edu/osr/faq.htm#gift, Accessed: 3 December 2004.

Darby, M. and L. Zucker (2002), 'Going public when you can in biotechnology', NBER Working Paper No. 8954, pp. 1–35.

Druilhe, C. and E. Garnsey (2004), 'Do academic spin-outs differ and does it matter?', *Journal of Technology Transfer*, **29** (3–4), 269–85.

Ernst & Young (2003), 'Queensland Biotechnology Report 2003', Brisbane: Ernst & Young.

Ernst & Young (2004), 'Resurgence: Global Biotechnology Report 2004', The Americas Perspective, Ernst & Young.

EVCA (European Venture Capital Association) (2004), Available: http://www. evca.com/html/home.asp, Accessed: 20 December 2004.

Genentech (2004), Available: http://www.gene.com/gene/about/, Accessed: 30 May 2004.

Haiduck, G. and S. Burrill (2004), 'Life sciences . . . circa 2004: biotech's back on track', conference presentation, Kellogg Biotech Conference, USA, 17 April.

Hamel, G. (1999), 'Bringing Silicon Valley inside', *Harvard Business Review*, Sept–Oct, 71–84.

Kerzner, H. (1998), *Project Management – A Systems Approach to Planning, Scheduling, and Controlling*, 6th edn, New York: John Wiley.

OECD (2001), *OECD Biotech Statistics*, September, Paris: OECD.

OECD (2002), 'The measurement of scientific and technological activities', Frascati Manual: Proposed Standard Practice for Surveys on Research and Experimental Development', Paris: OECD, Available: www.sourceoecd.org, Accessed: Dec 2003.

Project Management Institute, Standards Committee (1996), *A Guide to the Project Management Body of Knowledge*, Pennsylvania: Project Management Institute.

Scottish Enterprise Council (2004), http://www.ideasmart.org/index, accessed on 11 December 2004.

Turner, R. (1993), *The Handbook of Project-Based Management*, London: McGraw-Hill.

Vitale, M. (2004), *Commercialising Australian Biotechnology*, Sydney: Australian Business Foundation.

Zahra, S. and A. Nielsen (2002), 'Sources of capabilities, integration and technology commercialization', *Strategic Management Journal*, **23**, 377–98.

Zucker, L.G., M.R. Darby and J.S. Armstrong (2002), 'Commercialising knowledge: university science, knowledge capture, and firm performance in biotechnology', *Management Science*, **481**, 138–53.

4. Funding innovation in biotechnology companies

INTRODUCTION

There is evidence that innovation is more likely to occur in industries that are emerging, that are developing a track record of growth and success. This does not mean that innovation is solely the domain of young industries, but it is where major product innovations are more likely to be prevalent. Biotechnology is an emerging industry; however, it is also a capital-intensive industry (Ernst and Young 2001). A major issue for biotechnology is that due to high capital costs (particularly for research and development), technology diffusion will be slow without significant levels of investment. This investment tends to come either from large pharmaceutical companies or from public and private investors. Significant investment is required in the innovation process to ensure growth of these emerging industries.

Biotechnology has been recognized world wide as a critical sector to national economies, investment is crucial to the growth and continuity of the industry, as it is an industry uniquely characterized by high cost research and development (R&D), long-term but limited commercialization, and rapid change brought about by constant technological developments and scientific advances. This chapter provides an overview of how a biotechnology opportunity is taken to market. It begins by discussing how a biotechnology opportunity is assessed in its early stages, followed by a description of valuation techniques that can be applied to the intellectual property identified in the opportunity or early stage venture. Pathways to market for the opportunity are then considered and various funding sources are described.

OPPORTUNITY ASSESSMENT

This is an area where it is useful to know what is important to look for when assessing an opportunity, in anticipation of converting an opportunity into a successful venture. While you may not know all the pitfalls that might be facing the venture, it is important to recognize that there have been many ventures that have gone before. The astute founder will seek credible lists of possible flaws in related ventures, and hope to learn from other people's mistakes and as well as their successes.

Success and Failure

There are two aspects to the feasibility analysis for the biotechnology venture; those factors which can cause failure in a venture, and those positive factors that need to be sought out if a successful venture is envisaged.

According to Hatten (1997), the major cause of new venture mortality is related to economic/financial factors, in order of reported importance – insufficient capital, heavy operating expenses, burdensome debt, insufficient profits, industry/market weakness. These problems are generic to all new ventures. NBFs are not immune to these issues, in fact insufficient profits is a problem impacting almost all biotechnology companies. Further to this the usual issue of inexperience, particularly in the biotechnology industry in which most entrepreneurs are home-grown, with a strong science but an insufficient business background and experience have an impact on mortality rates. The final in Hatten's list, neglect, is rarely an issue in biotechnology given the extensive efforts required to establish the company from what is normally an extensive publicly funded research project. While not neglect per se, excessive concentration on one side of the venture, usually the technical/scientific may well lead to failure, linked usually to the failure to obtain financing.

In looking at the list we see a heavy orientation towards financial factors. They relate to liquidity, solvency and profitability, the three major areas of financial management for the venture. Certainly in biotech the last factor is less of a problem than in most industries. It is interesting however to see that inexperience is a relatively small part of the cause for failure in businesses. Possibly the reality is that inexperience leads to some of the other causal factors in the failure of businesses such as the financial factors.

Stages in the Feasibility Analysis

As with any industry-based analysis it is advisable to move from the macro considerations through the meso, and finally down to the micro considerations. That is, move from the broad industry assessments through to those factors which will specifically influence the success or failure of the specific venture.

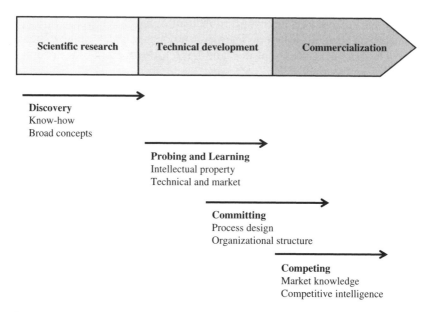

Figure 4.1   How emerging knowledge evolves

Figure 4.1 provides a pattern of knowledge development in an industry. Knowing where the industry, or more particularly the market niche, stands is important as the first step in screening a venture. This introduces the industry life cycle that will be discussed in further detail in Chapter 8.

It is difficult to find actual examples of feasibility studies undertaken by biotech companies. The list below provides examples of criteria that are relevant to the biotechnology industry at a macro level. The list provided below is pertinent to all biotech ventures as they consider their strengths and their future needs.

1. Access to business support services
2. High tech R&D base
3. High tech business base
4. High tech manufacturing base
5. Access to university and/or Federal laboratory technology
6. Access to infrastructure support
7. Workforce development programmes
8. Demonstrated access to seed/venture capital
9. Presence of 'champions' with a strong desire to drive the opportunity forward.

THE FEASIBILITY STUDY AND VENTURE SCREENING

Figure 4.2 displays a stylized map of the enterprise creation process. The starting point of every new venture is a problem, and an idea for solving it. The developmental process of turning the idea into reality requires an assessment at an early stage so that the entrepreneur can gauge whether the idea is worth carrying through to market.

Every idea is the product of human imagination (tacit knowledge), and many ideas are never progressed to the next stage (codified knowledge) where they can be evaluated or further developed. Ideas must be tested for fatal flaws before they can be set out as a concept. Concepts are a combination of an idea for resolving a problem and a practical proposal, in outline form at least, for delivering that solution. Many embryonic concepts suffer from fatal flaws, as discussed below.

A concept may address a real problem and be technically feasible, and yet it may be commercially unfeasible: the total cost of turning the concept into a marketed product, and then supplying the product to customers, may be too high in relation to the revenue that the new product would generate. The development of a business plan, is fundamentally about turning a concept into a commercially feasible proposal. Developing a complete business plan

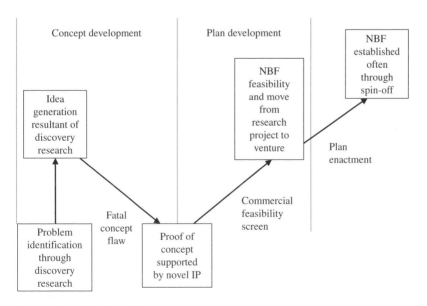

Figure 4.2   The feasibility process

takes time and effort, and it is disappointing, to say the least, to get to the end of the process and discover that the concept cannot be turned into a commercial proposition.

The following sections describes the process of turning an idea into a concept, and then applying a fairly simple commercial feasibility test that will screen out commercially impractical concepts before the full effort of developing a business plan commences.

Micro-level Analysis – the Technology Assessment

In assessing a technology a number of criteria need to be taken into consideration:

- Is the patent and literature search complete and clear?
- Is the ownership of the technology clear?
- Can the technology be protected?
- What is required to commercialize the technology successfully?
- Is this technology transferable?
- Will the technology give some sustainable competitive advantage?
- Are there competitive technologies?
- Does a definable and accessible market for the technology exist?
- How long will it take to get to market?
- Is a positive return on investment expected?
- Are product or application outcomes feasible?
- Can the technical risks be managed?

Only once the technology assessment is complete can a commitment to it occur. In the biotechnology sphere, where the technology is the core capability, the effective assessment of its feasibility creates the basis for the overall venture assessment. This also applies to the valuation of the technology or intellectual property. Hence the feasibility leads directly into the business plan.

OPPORTUNITY SCREENING CONSIDERATIONS – FATAL FLAWS

Fatal flaws will stop the progress of research and of the establishment of an NBF in its tracks. The following list of potential fatal flaws is not exhaustive, however, it can be used as a guide to minimize potential disappointment and heartache for people wanting to establish a new venture (Zimmerer and Scarborough 1996).

- **Scientific feasibility** The laws of science should not be contravened in the proof of principal stages of development. Hence there should be no contradiction of the scientific method, the central dogma theory of molecular biology, existing laws of physics, theories of relativity, the laws of thermodynamics, or the laws of conservation of energy and momentum.

- **Economic feasibility** While few laws exist in economics, it is advisable that principles and theories are not contravened. This is only an issue later in the commercialization process.

- **Technical feasibility** The existence of a patent does not guarantee a product can be feasibly created. There is an economic factor here as well, as the Pareto principle would dictate that excessive cost of new product development and later production, would preclude its market entry. Beyond a development point, production costs may escalate exponentially, to the extent that resource limitations are achieved before market launch is approached. Also a product can be scientifically and economically feasible without being practical given the current state of technology. This raises the technology push versus market pull development dilemma. In some areas of biotechnology it is more a matter of the final customer needing it rather than wanting it where a therapeutic component is involved.

- **Marketing feasibility** Put simply: does anyone want it, or does anyone need it? Has the product any features that would persuade someone to choose it ahead of currently available products? The Biota example is an obvious one here.

- **Legal/regulatory** Biotechnology is a highly regulated industry. If a lead candidate is likely to kill or cause human harm, it will not reach market (with the exception of Vioxx and a few other drugs).

VALUING THE OPPORTUNITY

Once the fatal flaws test is complete, the feasibility of establishing the NBF can be considered, as the NBF is the commercial vehicle which will take the candidate from the research lab through to the market, generally with a lot of help from governments, allies and larger biotechnology and pharmaceutical companies. The longer term success of the NBF will be determined by some critical issues, central among which are:

- Final selling price of the product is a major issue in valuing the IP of the company and the valuation of the company itself, even though most NBFs will never come face to face with a final consumer in their existence.

- Most business plans for NBFs state how many million people around the world suffer from a certain affliction and interpret this as their market. Understanding the real market segmentation, their target market, their micro-market is what is critically important rather than a mythical world-wide market.
- Variable unit costs of production are an issue, though not so much for NBFs as they are unlikely to be involved in manufacturing the final product.
- Fixed costs are critical in the biopharmaceutical area as the regulatory requirements including all the clinical trials present significant intractable costs in NPD.
- Capital requirements for the NBF are a never-ending issue in a capital-intensive high-technology industry. The lease or buy decision is pertinent on much capital equipment, as well as for the location of the company.
- Rate of return is an important issue and one we discuss further in this chapter.

The economic profit, if any, is given by the formula:

$$Profit = units \times (selling\ price - variable\ unit\ cost) \\ - fixed\ costs - rate\ of\ return \times capital$$

For many biotechnology ventures the concern about selling prices and unit sales is foreign or too early to be considered during the discovery stage of a long development pipeline. The value of the biotechnology venture is therefore related to the exit strategy or when the intent to divest is likely to occur in the life cycle of the venture.

For many biotech ventures the customer is another company. For many their relationship with final consumers is limited. Therefore the feasibility analysis and the business plan have to consider the audience which will be interested in the product (the same can be said of the IP, where the user should be considered in drafting the patent documents).

What is the value of an opportunity? The question applies equally to the venture as it does to a final product. Both are linked strongly to the valuation of the IP, especially when the venture is in its start-up phase. The following section outlines a number of valuation techniques and methodologies for early stage ventures or IP.

Valuation Techniques

Valuation of IP or early stage ventures continues to remain one of the most difficult and subjective tasks in the biotechnology industry. The valuation of an opportunity is linked to funding. Because of the long time-frames associated with getting a product to market most of the value in IP is associated with the future benefits that will flow from the IP or venture. It is important to note that the ultimate value of the IP is the price that someone is prepared to pay. This section presents an overview of the different valuation methodologies available.

A fundamental difference exists between the value and price of IP; terms that are commonly interchanged. Value represents the future benefits from IP ownership and is related to the benefits to the IP user. Price is the amount at which the IP would exchange ownership between a willing buyer and a willing seller, neither being under compulsion to deal, with benefits accruing to both. Value is not necessarily equivalent to price. In simple terms, value is based on an opinion resulting in different answers based on the valuation technique used, while price is an outcome or commitment of a negotiation that has taken place.

A number of reasons why a valuation is required include (McGinness 2003):

1. To assist in determining licensing royalties and negotiating deal parameters;
2. When securing financial investment in the venture;
3. As a basis for establishing potential damages for IP infringement;
4. For legal and accounting standards requirements; and
5. For taxation, particularly capital gains tax and stamp duty liabilities.

The value of an opportunity or technology is established at a specific point in time and can be done prospectively or retrospectively. It is done prospectively by deal makers entering a negotiation. This includes willing buyers and sellers, investors, lenders and acquirers. When a valuation is done retrospectively it is usually by litigators where a judicial outcome is imposed as a result of an adversarial incident, i.e. unwilling buyers or sellers exist.

A common situation where IP valuation is undertaken is during a technology licensing deal. Technology licensing involves the following components:

- Technology rights
- Risk
- Art of deal making
- Deal parameters and economics.

The value matrix	Precedent	Paradigm	Possibility
Price	Industry standards/ benchmark	Auction methods	Real options methods
Worth	Rating/ranking method	Rule-of-thumb	Monte Carlo methods
Compensation	Rule-of-thumb	Discounted cash flow	Net present value

Source: AUTM 2004

Figure 4.3   The value matrix

Different valuation methods are oriented toward different perspectives. The value matrix in Figure 4.3 shows the common valuation methods, from different perspectives and when they are likely to be used.

Generally, there are three broad valuation approaches commonly used: the cost-based approach, the market approach and the future income/revenue approach. However, a number of methodologies exist to provide some indicator of value. It is important to note that the different methodologies will produce different valuation results and some methodologies will be more appropriate than others depending on a particular situation. Ultimately, the true value of IP is determined by the particular price someone is willing to pay. The following section describes the most common valuation methodologies in more detail.

Cost-based

Using the cost-based approach, the value of the IP or associated technology is assumed to be equivalent to the historical cost of development and protection. Usually, a return is added to the base costs. It is important to ask if the cost to develop the IP or technology is relevant to the situation. The following areas are relevant where cost enters a licence negotiation:

For academic institutions:

● Sunk patent costs
● Modified replacement cost method used upfront

– The Modified replacement cost asks what it would cost you now to redevelop the technology or IP, knowing what you have learnt. The value is usually lower than the original cost.

For corporations:

• This method is used when the transfer of know-how is involved.

It is recommended that the cost approach is used for legal and accounting standards requirements or when comparable market information is not available (Sullivan 1994). It is also useful during litigation when damages from IP infringements need to be determined.

Industry Standards/Comparable Market

The industry standards/comparable market approach measures the present value of future benefits by obtaining consensus from other similar transactions between unrelated parties that have occurred in the marketplace. This is the most common approach used in the licensing world since it reflects a fair market value for the technology or IP. If a comparable or market benchmark exists then this is the preferred method as this is a credible valuation technique. A number of sources are available to identify comparable transactions:

• Internal database – licences previously worked on by own organization
• Published surveys – relatively very few in number
• Public announcements – Internet searches and database services
• Word of mouth – contacts in the industry involved in deal making
• Litigation – outcomes and documents from trials
• Required disclosure – for example those contained in securities and investment commission documents in a number of countries.

Some disadvantages exist with this methodology. First, an active market with similar technologies must exist. Second, this method assumes that current industry norms are correct, and finally, a cost is involved to obtain accurate data from previous industry deals.

Rule-of-Thumb (25 per cent Rule)

This method values intellectual property by calculating a royalty of 25 per cent of the expected gross profit, before taxes, from the enterprise operations in which the intellectual property is used (Sullivan 1994). This method, while it is a rough estimate, is useful to generate a ballpark valuation. It is also useful when the intellectual property is at a commercial stage or for litigation purposes. It should always be used with caution since it does not consider a fair return on investment nor potential profitability.

Discounted Cash Flow/Net Present Value

The value of property can be measured by the present worth of the net economic benefits (cash receipts less cash outlays) to be received over the life of the property (Sullivan 1994). The discounted cash flow is the most accepted method in the financial marketplace since the majority of people are comfortable and familiar with the time-value of money. Using this methodology requires compensation for the following:

- Inflation
- Risk
- A return on the investment.

This methodology can be applied at any stage of development of the technology; only the ability to obtain accurate data will limit this approach. This approach also requires significant knowledge of the competitive environment.

Rating/Ranking

The rating/ranking method compares the intellectual property asset in question to comparable intellectual property assets according to a subjective scale, using panels of experts, or to an objective scale based on measurable past experience. The five elements that comprise the rating/ranking method are as follows:

1. Scoring criteria
2. Scoring system
3. Scoring scale
4. Weighting factors
5. Decision table.

This approach influences people to prepare for negotiation by thinking through the relevant factors that make up licensing value. It facilitates discussions with other valuation experts as it focuses on the key components of value in the technology.

Monte Carlo

The Monte Carlo method provides approximate solutions to a variety of mathematical problems, including valuations, by performing statistical sampling experiments on a computer. The method is useful for obtaining numerical solutions to problems which are too complicated to solve analytically. The Monte Carlo method is based on the discounted cash flow approach but incorporates uncertainty to provide a more rigorous analysis. This is done by assigning a range of values to the variables used in calculating the net present value of an asset. The probabilities calculated provide greater insight and can identify which assumptions drive overall uncertainty. This approach is more complicated and harder to understand than other valuation methods. It is also difficult to estimate and obtain agreement on probability distributions.

Auction

The auction approach discloses information on the IP or technology to broad potential customers and accepts different sealed bids (Sullivan 1994). This approach is useful when a number of interested buyers exist or when up-front payments are being targeted. No calculation of valuation is required using this approach as it assumes that the highest bid will reflect the true market value. The disadvantage with this method is that the willing seller has less control in setting the price as this is left to competitive market forces.

Real Options

Real options analysis is designed to explicitly incorporate and analyse risk and uncertainty associated with real assets. The real options method applies financial options theory to quantify the value of intellectual property. It views an opportunity as a process that managers can continually reshape in light of technological or market changes. Because the real options approach recognizes that risks can be managed, to avoid bad outcomes or take advantage of good ones as they become apparent, the use of real options practically always leads to higher values for the same project than the traditional methods. This is precisely because the options perspective recognizes that managers make future decisions about a project as uncertainties become resolved. The downside to real options valuation is that it can be a difficult and time-consuming process.

It is recommended that where possible, a number of the methods outlined above be used. In addition, any available industry standard or benchmark provides a level of objectivity that is easier to defend in negotiations than projected revenues. The exception to this rule is where the market is under-valuing the type of IP, from the company's perspective, then evidence must be provided to support this claim.

ALTERNATIVE COMMERCIALIZATION OPTIONS AND PATHWAYS TO MARKET

Spin-off

Spinning off the technology or IP into a start-up allows focus on commer-cializing the IP. Advantages include access to government assistance pro-grammes or pre-seed funds. Disadvantages include requirement for a significant capital outlay and resources, risk associated with medium to long time-to-market (especially biotechnology/pharmaceutical projects) and the inherent risk in establishing a new venture.

Traditional out-licensing of technology or IP

Licensing involves the transfer of rights to make, use or manufacture patented IP. It is a low-risk alternative to the spin-off option requiring low capital outlay. However, returns are also comparatively low depending on the progress of the technology along the value chain. When the technology has progressed to prototype stage or clinical trials then a higher royalty rate can be obtained. Negotiations usually involve an upfront licence fee with annual royalties usually payable per quarter. They can be subject to milestones set by the licensee and agreed by the licensor. A minimum annual fee is set for royalty payments. If royalty payments are lower than the minimum annual fee then the difference is payable by the licensee. Licensing involves transfer-ring the rights to the technology/IP on an exclusive or non-exclusive basis. Licenses also define fields of use, geographic areas, economic limits and pro-duction and distribution limits. A licence agreement stipulates the licensing arrangements including the licensing fees and royalty payments.

Trade sale

The rights to a technology or IP are transferred 100 per cent exclusively to a client in exchange for a single payment or a series of progress payments. This approach allows IP to be divested or transferred to an entity that per-ceives greater value in the IP or is readily able to exploit the IP. This allows the IP provider to redirect the payments to fund the further development of higher-value technology or IP. The only disadvantage with this approach is the loss of total control over the IP and the emotional attachment that may still exist.

Internal development

Technology or IP is further developed internally either as a technology platform, converted to a product or as a hybrid where the technology platform is used to further develop products or applications. This approach adds value to the technology/IP as it is further progressed along the value chain. This approach implies that the organization has the resources and capability to progress the technology or IP across the full spectrum of the value chain; from discovery, through development and finally launched to market. The internal development option is limited to organizations that have established capability to achieve this, including research, development and prototyping, clinical trials and regulatory approval, manufacturing, and sales and marketing. Examples of biotechnology organizations that possess this capability include, Genentech, Amgen and Abbott Laboratories.

IP bundling/packaging

Grouping of technologies or different IP with synergistic or complementary characteristics where the sum of the technologies is far greater in value than the individuals. Once the technologies or IP is bundled, options 1 to 4 above can be used to commercialize it.

Incubation

The technology/IP is incubated within another institute, organization or cooperative research centre where the technology is further developed. The advantage of this is using resources within the other organization to further develop this technology. The disadvantage of this approach is clarifying the ownership of existing IP, further development of the IP and any new IP that is generated. This approach may involve joint ownership by all parties where the ownership is clearly articulated in an agreement by prior arrangement. The agreement may also specify joint equity based on the contribution by all parties involved.

Collaboration/partnerships/joint venture

This is similar to the incubation option, however, specific partners or collaborators may be used at various stages of the development of the IP. Collaborations are usually established at an early stage while a joint venture may be formed at a latter stage when the technology or product is more defined.

Stepping-stone approach

A strategy for commercializing the technology/IP is mapped along the various stages of the value chain and may involve a combination of internal development, licensing and collaborations. This type of approach is quite complex and may involve through licensing across the different stages of the value chain.

Simple contract
IP or know-how is simply transferred and applied to an organization via a simple contractual arrangement. This may be based on a single or multiple fee structure payable upfront, progressively or as a progress payment after the transfer has successfully occurred. This may also involve royalty payments but is not common. Many contract R&D, product development, manufacturing, marketing and distribution agreements are executed using this option.

COMMON COMMERCIALIZATION OPTIONS

Commercialization of research and knowledge requires a number of decisions to be made. The flowchart in Figure 4.4 summarizes at a high level the common commercialization options practised by technology managers.

A second major decision is to identify the point at which the technology or IP is harvested along the value chain. Greater value is gained as the technology or IP is progressed along the value chain. See Figure 4.5.

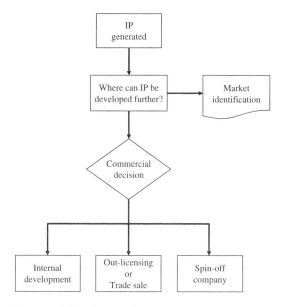

Figure 4.4   Commercialization options

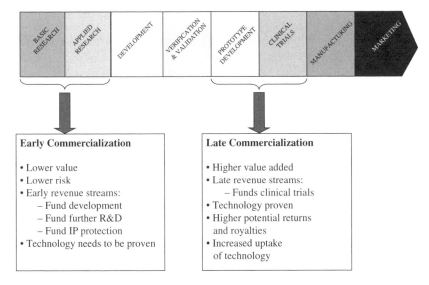

Figure 4.5   Early and late stage Commercialization

Forms of Technology Transfer

Figure 4.6 can assist in the decision-making process when considering a technology transfer.

ENTRY STRATEGIES AND BARRIERS TO ENTRY

Entry strategies can stem from an analysis of the stage of life cycle in which the market currently functions. Some of the possible entry strategies for new ventures are listed below, however the list is not exhaustive.

New product or service
These will be rare and sometimes difficult to imitate, though this is not necessarily sustainable as Biota and Resmed have recently found out, especially if the product is easily imitable when the patent runs out.

Parallel competition
This involves introducing competitive duplications onto a market – particularly generic drugs after patent expiry or accessing public domain gene sequences. The duplications are parallel though not identical to the existing product.

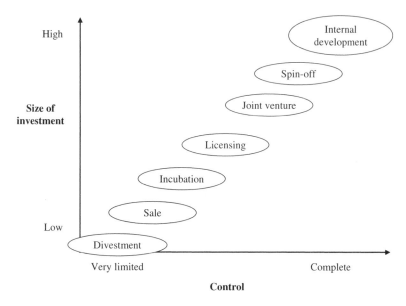

Figure 4.6   Forms of technology transfer

Franchising

This takes a proven formula for success and expands it. It means the fran-
chisor can expand rapidly without the risk of excessive capital outlay.
Unfortunately for the franchisee the capital burden is borne more by them.
Further franchises tend to be established in the mature phase of a PLC and
therefore when the franchisee is locked into a single standardized product
their life will follow the path of that product, which may be rather short.
Franchising allows small firms to exist where small firms probably should
not exist through achieving economies of scale and scope. There isn't much
evidence of franchising around biotechnology yet, but areas such as contract
lab research and clinical research may lend themselves to this in future.

Barriers to Entry

The following are some examples of barriers to entry which may restrict the
emergence of new ventures in markets.

Retaliatory barriers

These are anti-competitive actions as a result of the entry of a new competitor in a market. This will occur where there are larger firms which have
control over price or quantity supplied in the market (they have some
degree of monopoly power).

Regulatory and statutory/legal barriers
These may include patents, Acts of Parliament, regulated markets common in the biotechnology industry, trademarks, copyright and other intellectual property.

Structural barriers
These are based upon economies of scale, capital outlays, set-up costs and average production costs in a price competitive market.

One of the fundamental considerations is the stage of development of the market. This can be judged in various ways, however the important consideration is the extent of impact of economies of scale.

Economies of scale refer to the reduction of long-run average costs for businesses. This is usually achieved by an increase in volume of production of their goods and services which has to be underpinned by increasing demand for their product. As volumes of production increase, purchasing power increases for the business allowing it to purchase its inputs more cheaply. It also begins to use its resources more efficiently and therefore reduces its costs of production accordingly. This increased efficiency of use of its resources mean that it can expand production and starts the process of becoming a larger business. It can lower its prices by having lower costs and therefore begins to take a greater share of its market and the market itself changes to one in which price competition becomes important.

It is in this situation that the larger businesses become more competitive and smaller businesses, because of their higher cost structures, become less competitive. Therefore, in general, in industries where economies of scale are starting to impact and price competition is becoming important, small businesses have little competitive advantage. This is where the establishment of a new business becomes more problematic.

WHERE DO THE INVESTMENT FUNDS COME FROM?

When we consider new ventures, there are obviously two types:

1. The start-up, usually based on new IP and being undertaken as an independent business;
2. The establishment of a new venture by an existing business, either as a semi-independent entity (such as in a joint venture) or a separate business unit.

The situation for each of these two venture types will differ dramatically in terms of funding sources, management, staffing and administrative, and accounting systems, as well as legal standing. For the start-up, life will be far more difficult due to limited availability of early-stage risk capital. Figure 3.4, provided details of sources of equity financing available to NBFs.

What makes the biotechnology industry unique, especially in the bio-pharmaceutical sector, is the amount of funding required to start a biotech company and even the greater subsequent funding required to sustain the growth of the company. A typical biopharmaceutical company requires about 2–3 million dollars in its first two years and between 5–10 million dollars in its second two years. After this period funding of up to 600 million dollars may be required to progress a new drug through clinical trials and finally to market release. Venture capital (VC) funding is therefore an important driver of the biotechnology industry.

Venture Capital

Typically VC funds fall into two distinctive areas.

- The formal venture market – largely banks, financial institutions and other institutions providing debt financing both short and long-term to businesses.
- The informal venture market mainly made up of two types of venture capitalists: Family and friends – this is common in many business practices, especially to establishing start-up business.

A sub-set of the informal venture capital markets are the business angels: who are private individuals also known as informal investors who provide risk capital directly to new and growing businesses in which they have no family connection.

The informal venture market is considered to be significantly larger than the formal market and that business angels fill the so-called equity gap by making investments precisely in those areas in which institutional venture capital providers are reluctant to invest.

Types of venture capital available
Finance requirements are based upon two major factors:

- The stage of life cycle of the firm – Start-up firms find it difficult to obtain funding from the traditional sources, however venture funding may be forthcoming in the form of equity funding from a business angel.

- The growth aspirations of the firm and its track record in this area – Those firms that have a track record if they are high growth, may also have amassed some tangible assets which will make long-term debt financing more achievable.

Preferences Between Debt and Equity Financing

Debt financing

This requires the venture to pay back loans with interest rates often higher than that charged to larger firms due to the risk factor in the lending. Debt financing does however allow the owner/management of the venture to maintain control of the business.

Equity financing

This allows the business to grow without a significant debt burden to slow its progress and limit its cash flow. However equity capital usually comes at a price. Venture capitalists often insist that they have significant voting rights in the company and may further insist that they take over the running of the company while the original owner is relegated to a technical position. The degree of intervention of the venture capitalist in the running of the business is dependent upon the degree of development of the business concept.

It must be remembered that profit is the motive for most venture capitalists as they are providing capital to firms which could not obtain funding from other sources, therefore they need to control their risk by ensuring that the management aspects of the business are in their control. They are effectively then controlling the internal environment of the organization so that they can better deal with the turbulent external environment faced by many of the ventures.

THE EFFECT OF ALTERNATIVE BUSINESS MODELS

The first modern biotechnology companies (e.g. Genentech, Amgen) adopted a fully integrated pharmaceutical business model. The attraction to this vertically integrated model was the ability to manage and control the full spectrum of the value chain. The downside was that these companies were also exposed to the maximum level of risk associated with the financial capital requirements to maintain this infrastructure.

Existing biotechnology business models have evolved to meet market needs and deliver returns on investment. Fisken and Rutherford (2002) identified three different business models in European biotechnology companies, based largely on financial and economic criteria.

The Product Business Model

The product business model originates from the pharmaceutical model where value is added along the activities of the value chain to deliver a final product to market. This is a proven model where 19 of the top 22 biotechnology companies in the world with market capitalizations in excess of US$3 billion are classed as product companies.

The Platform Business Model

The platform business model is a relatively new concept. This model aims to generate value from the front end of the industry value chain. The model is based on the development of research tools or platform technology that can provide a service to another organization or can be licensed for further development along the value chain. The evolution of this platform originated from technological advances applied to drug discovery and was driven by the need to reduce risks and minimize financial capital requirements. Due to increased levels of technological change, the platform model is no longer viewed as low risk. Many companies are now moving to a hybrid business model.

The Hybrid Business Model

This business model is a hybrid of the product and platform business model. It generally consists of a platform technology that is able to generate a pipeline of products. This model is an evolution of the platform business model in that it aims to generate value through downstream activities of the biotechnology value chain. The pipeline of products can be developed organically or through additional in-licensing, purchasing exclusive or semi-exclusive access to another's technology, to enhance product development opportunities.

The convergence of the product and platform business model to create a hybrid model provides investors with the benefit of reduced risk and the potential for future revenue growth. This approach will need to become the preferred model for new and existing biotechnology companies to maintain growth and sustainability in the industry.

The three different models are outlined in Figure 4.7.

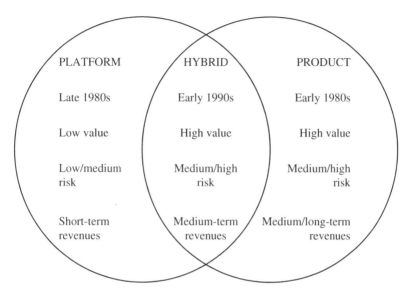

Source: Adapted from Fisken and Rutherford 2002

Figure 4.7   Characteristics of platform, product and hybrid business models

CONCLUSION

The chapter outlined how a biotechnology opportunity is taken to market and how it is assessed in its early stages. Different valuation techniques for intellectual property were discussed. Pathways to market for biotechnology opportunities were considered and various funding sources were described. The chapter also looked at the alternative business models in biotechnology.

REFERENCES

Association of University Technology Managers (2004), *Technology Transfer Manual*, University of California, Irvine.
Ernst & Young (2001), *Australian Biotechnology Report*, Ernst & Young, Freehills and ISR, Commonwealth Department of Industry, Canberra, Science and Resources.
Fisken, J. and J. Rutherford (2002), 'Business models and investment trends in the biotechnology industry in Europe', *Journal of Commercial Biotechnology*, **8** (3), 191–9.
Hatten, T. (1997), *Small Business Entrepreneurship and Beyond*, Englewood Cliffs, NJ: Prentice Hall.

McGinness, P. (2003), *Intellectual Property Commercialisation: A Business Manager's Companion*, Lexis Nexis Butterworths.

Sullivan, P. (1994), *Profiting from Intellectual Capital – Extracting Value from Innovation*, New York: Wiley.

Zimmerer, T. and N. Scarborough (1996), *New Venture Creation*, New Jersey, Houghton Mifflin.

5. Intellectual assets I – intellectual capital in biotechnology firms

INTRODUCTION

This chapter discusses the underlying crucial importance of intellectual capital to the biotechnology industry, in the context of absorptive capacity, tacit and codified knowledge and IP, innovation and creativity. Intellectual capital is vitally important to the successful biotechnology organization. However knowledge sharing is a dilemma in an industry torn between the disclosure inherent in the scientific process, particularly through publication, and the need to maintain confidentiality and control over IP. Therefore simplistic measures such as the number of patents held by a company, or the calibre of its research staff, cannot measure intellectual capital; intellectual capital is only an asset when it is realized, through sharing unique knowledge held by staff to create unique products. Patents provide some evidence of this uniqueness, but are only one outcome of intellectual capital. It is an area that requires adept management and strategy.

Alternative patent strategies, particularly where used as defensive strategies to limit competition are employed to great effect by larger pharmaceutical and chemical companies. Bioentrepreneurs have to be acutely aware of such strategies as they consider their own IP strategies. Patents normally provide protection for 20–25 years, however the average drug takes at least 15 years for completion of development, testing and regulatory approval. Furthermore, patent protection is usually sought very early in the R&D process. Companies therefore have only a short timespan to achieve returns on their investment before patent protection expires and competitor products can be legally launched on the market. Substantial returns are needed to cover the high development costs. Meanwhile in order to have their products market ready when IP protection expires, competitors are using the patent documents and publications in scientific journals to access the codified knowledge required to replicate the product.

Journals important in this research include *Nature, Lancet,* and other more specific high citation impact journals such as: *Trends in Biotechnology, Pharmacogenetics, Journal of Computational Biology, Genome Research, Bioinformatics, Applied Environmental Microbiology.* In this chapter the importance of ownership of intellectual property, creating effective IP strategies and their impact on the valuation of the young biotechnology venture are analysed.

A firm's intellectual assets can exist in a variety of forms. In some instances, knowledge can be codified in a formalized or systematized manner. In practice this means that the knowledge has been transcribed and explicitly defined, the most common variant of this form is intellectual property. By definition, intellectual property protection can only be granted over codified knowledge.

Other forms of intellectual assets are described as tacit knowledge, which is acquired through 'learning by doing'. Tacit knowledge can only be internalized by a firm through practice. This knowledge cannot be easily depicted by words and in some cases cannot be verbalized at all. Tacit knowledge encompasses skills and experience.

Intellectual capital (IC) must be considered central to the competitive advantage of the growing biotech company. The company cannot rely on existing IP alone and must generate new IP if it is to sustain its growth. In this regard, IC goes beyond human resources. While human resources will have regard for what is considered to be the social capital of the company, intellectual capital is a more specific concept, it is the intellectual resource of the company. Not only should the people in the company be nurtured, but their intellectual capacity must be appreciated and the sharing of their knowledge facilitated. Knowledge sharing will only occur when staff consider their ideas are listened to and they will gain some reward for their intellectual efforts. This may include recognition in patents, promotion, increased research funding and staff. Intellectual capital, as it is held in people's heads, is a very transitory resource. For such assets to be maintained over the long term, not only do good staff with unique knowledge need to be retained, the sharing and diffusion of their knowledge within the company must be achieved. Until the knowledge is shared it cannot be considered an asset.

For biotechnology organizations, their advantage is seen to be gained by their Intellectual property. It is this which is measured on the balance sheet of the company. However despite what the lawyers would want us to believe, there is something far more vital to the biotechnology company if it wishes to grow. It is not just the people they employ or work with, but the potential of those people to create IP in the future. This intellectual ability residing within the company is often referred to as intellectual capital. Without intellectual capital, the gains made for the company with its IP cannot be sustained. For a growing company securing the future is a never-ending battle. It is only through innovation – product, process and organizational that this future can be realized.

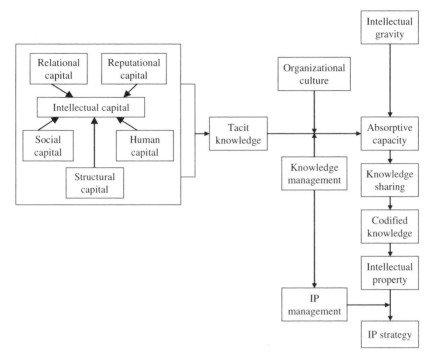

Figure 5.1   From intellectual inputs to intellectual outputs

There is a lot of terminology surrounding this area. Terms such as tacit knowledge, codified knowledge, absorptive capacity, embeddedness, radical, incremental, architectural and modular innovations and so on are introduced and explained on the way to exploring this important area of innovation and entrepreneurship.

Figure 5.1 seeks to bring together many of the important elements of knowledge and intellectual assets of the NBF which have the potential to make it and keep it competitive. Nowhere is the term 'organizational learning' used, instead concepts that have been explored and tested empirically are employed to construct the nature of the transformation process from intellectual inputs to intellectual outputs. This chapter and the next will cover each of the major aspects of this transformation process. The process is also presented as sequential so as to understand the relationships and some of the causality in the transformation.

While, as in many of the other areas of this book, the concepts are not specific to biotechnology, the analysis of intellectual assets (which combines intellectual capital, intellectual gravity and intellectual property) is absolutely critical to this industry, as it is so dependent upon the establishment, protection and maintenance of its intellectual assets for its competitiveness. Each organization which exists in the industry, is subject to this regime which pivots off the patent system of IP protection. Intellectual assets then are fundamental to the operation of any firm in the industry and essential to the establishment and existence of an NBF. As we will explore in more depth in the analysis in Chapter 6 on product life cycles, for sustainable competitive advantage a 'one-trick-pony' is insufficient. The NBF must be able to move beyond the single product or research project. It needs to establish a research programme, in effect an innovation stream which has the potential to generate a product stream. A single patent may achieve a degree of competitive advantage for the NBF to bring itself into existence. However, it will need a well constructed patent portfolio and beyond this, a well planned IP strategy to build long-term viability. To achieve this requires not only money, but people: people with ideas, who are willing to share these ideas and management which can look after these people to the extent that they remain with the organization and feel inclined to share their ideas within the firm.

Figure 5.2 sets the scene for a wide-ranging structured analysis of this enormous area which can be brought under the auspices of intellectual assets.

INTELLECTUAL CAPITAL

Intellectual capital is the sum of ideas and capacity to generate ideas in the people involved in the company, at its heart lies in Polanyi's (1962) concept of tacit knowledge.

Intellectual capital has been defined as 'the sum of everything everybody in a company knows that gives it a competitive edge . . . Intellectual Capital is the intellectual material that has been formalized, captured and leveraged to create wealth' (Stewart 1997). A firm's intellectual capital represents the environment out of which patents and other intellectual property can be generated. It includes the firm's absorptive capacity: its ability to exploit information across the organization. A crucial difficulty with the intellectual capital construct lies in its measurement and valuation.

While there has been much research in the past decade attempting to delimit the concept of intellectual capital, the current consensus appears to define the intellectual capital construct in terms of three components:

- Relational (external, for example, consumer-related) capital;
- Structural (internal or organizational) capital; and

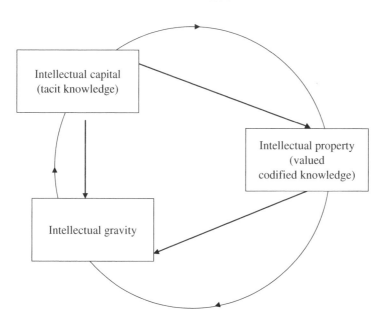

Figure 5.2   The relationship between the major concepts raised in these chapters

- Human capital (Palacios-Marques and Garrigós-Simón 2003; Bontis et al. 2000; Roos et al. 1997).

Relational capital refers to all knowledge assets amassed by the firm from its relationships with competitors, suppliers, industry associations, government and other relevant organizations (Bontis 1998; Bontis et al. 1999, 2000). *Structural capital* comprises the elements of organizational processes and activities that are linked to the creation of company value (Carroll and Tansey 2000). It is comprised of knowledge assets which have been systematized or made explicit by the firm, such as intellectual property; the knowledge embodied in infrastructure such as tools and equipment for research, measurement, and manufacturing; and other internalized knowledge shared informally inside the firm. *Human capital* may be described as the collective capability of one firm to extract the best solutions from the knowledge of its employees (Bontis 1998). It comprises all the tacit and explicit knowledge assets residing in individual employees – the know-how, information, relationships, and general capabilities that employees bring to the working environment.

Human capital therefore consists, not only of the technical and professional qualifications and experience of employees, but also of factors impacting on their relationships with other employees and customers – such as the ability to cooperate, lead, share and build knowledge; communication capacity; ability and willingness to collaborate; and the ability to adapt to difficult situations, to negotiate and to accept diverse opinions, and to act based on confidence of success (Palacios-Marques and Garrigós-Simón 2003).

It is critical for the biotechnology company to convert intellectual capital into outcomes. The outcomes of intellectual capital are largely, in the biotechnology company, intellectual property. As a capital resource such as physical capital, human capital, reputational capital and social capital, it should be noted that just because the resource exists, doesn't mean it is fully available for utilization or exploitation. In this case intellectual capital as a resource, resides in people's heads. Therefore this resource needs to be supported and nurtured so as to capture the full benefit of its potential.

The outcome of intellectual capital is intellectual property. However there appears to be, in many biotechnology companies, an overemphasis on intellectual property. Intellectual property is only the output of intellectual capital yet it provides a competitive advantage for the company, albeit a short-to-medium-term advantage. Intellectual property is the historical legacy of intellectual capital. However intellectual capital is more about the potential to generate future ideas within the company based on the intellectual capacity, ability to collaborate, and propensity to absorb new knowledge of the people within the company and associated with the company. It is through intellectual capital that the NBF can maintain a competitive advantage for the company into the future. It is therefore imperative that this resource, the people in the company, are supported. Unfortunately for many companies, intellectual capital is probably the most transient resource available to the company.

MEASUREMENT ISSUES

Importantly, in the sense clearly enunciated in the resource-based view of the firm, intellectual capital is inimitable (Barney 1991). This makes it a valuable, albeit transient resource. Zucker et al. (1998) further emphasized the impact of intellectual human capital in the US biotechnology industry. In their paper based on data from 751 US biotechnology companies and 327 star scientists, they concluded that the growth and diffusion of intellectual human capital was the main determinant of where and when US biotechnology developed.

Intellectual capital is no doubt important to the successful biotech organization. It cannot be assessed by the number of patents held, or even how bright the research staff may be. Intellectual capital is only an asset when it is realized, through sharing the unique knowledge held by staff to create unique products. Patents provide the evidence of this uniqueness, but are only the outcome of intellectual capital. It is an area which requires adept management. The management process was analysed in Chapter 3, Innovation and R&D Management.

Intellectual capital while critically important, is bedeviled by its transience and the difficulty of linking it with definitive innovative outcomes. The problem then with obtaining market value from intellectual capital is that the value from these assets is indirect. As not only an indirect influence, but an intangible one it is difficult to clearly measure the value and impact of this resource. The problem has so overwhelmed the accounting profession that international standards are reducing coverage of intangible assets rather than increasing them due to the inherent fuzziness and qualitative nature of measures. The accountants have opted for the obvious and the measurable and maintained their focus on tangible assets rather than assets whose cause and effect are ambiguous.

It is anecdotally clear though that a firm that captures and utilises unique knowledge capabilities effectively will tend to attract other experts and high quality professional staff thereby achieving increasing returns to scale. Measuring this through net present value and real options methods is, however, in its empirical infancy. The core activity of the firms which achieve increasing returns is research. It is the generation of new knowledge that leads to products with commercial value and return. The strategic goal of the firm is to establish a stream of innovations, each capitalizing on the success of its predecessor. Intellectual capital is thus the primary source of wealth creation since it enables the generation of new knowledge within the firm to establish and maintain technological leadership (Carayannis and Alexander 1999).

Palacios-Marques and Garrigós-Simón (2003) have operationalized the intellectual capital construct in terms of a multi-dimensional scale of measurement, subsequently validated by surveying managers from a large number of biotech firms. Based on their attempts to assign a relative weighting to the three components of intellectual capital described above, these researchers concluded that for biotech firms, human capital is the most important component, followed by relational capital, and finally structural capital.

An accurate reporting of intellectual capital will rely on the convergence of management accounting and financial accounting. Management accounting traditionally provides internal information and is primarily used for a firm's decision making. While financial accounting supplies external information and is considered as the main indicator for the company's value.

INTELLECTUAL CAPITAL IN THE WINE INDUSTRY

Research currently being conducted on innovation diffusion in the wine industry (Hine and King 2001) has found that much of the intellectual capital in the industry is vested in the winemakers. These winemakers operate in three distinct ways:

1. They can be the owner/winemaker of a winery (usually small);
2. They can be employed by the medium to large wineries; or
3. They can work on a contract basis for a number of wineries – often in different wine-growing regions.

For those working on a contract basis they share their ideas in different regions but the art of winemaking is vested in them as individuals. The wineries they service cannot consider the intellectual capital made available by their contract winemaker as an intangible asset of the firm, nor can they codify and institutionalize the knowledge by converting the art of wine-making and the creative idea generation of the winemaker to intellectual property, except to the extent that their work has influenced the wine's quality and features. In effect the intellectual capital lies outside the organization while still residing within the industry.

However for the other two types of winemakers both the intellectual property and the intellectual capital reside within the firm for as long as that person remains with it. In essence the intellectual capital of these wine-makers, when it resides within the firm for more than one accounting period, could be considered as a recurrent intangible asset. The intellectual capital made available to wineries by the contract winemaker could not be regarded as an asset as it does not remain within the firm for an extended period. One of the fundamental definitions of an asset is that it must be under the control of the entity (Martin and McIntyre 1996).

It must be remembered then that intellectual property is the output of intellectual capital, just as innovation is the output of creative management and the production processes. As an output it cannot deliver competitive advantage. To remain competitive there must be an ongoing stream of intellectual property out-putted. The firm therefore depends on its intellectual capital to be maintained in order to create the competitive advantage it requires. Intellectual capital being vested in individuals creates a dilemma for the firm. It becomes an imperative then for the firm to institutionalize the intellectual capital, to either remove it from the individual or to create a long-term relationship between the individual and the firm.

The subjects of this study were the winemakers/owners and in some instances both the winemaker and the owner were interviewed (where these were different individuals). The research found that the network can be an effective vehicle for the diffusion of innovations through the dissemination of ideas. For example the contract winemakers offer their creativity and expertise. However, it is a two-way street. In return they get to work with new wineries which are willing to experiment with new ideas such as grape varieties, blends and maturation techniques. The more established, older wineries are less willing to experiment and will concentrate more on wine consistency as a measure of quality. The opportunity to innovate is the element provided by the small wineries in South East Queensland which creates the mutually beneficial relationship which attracts the contract winemakers to that region.

As artisans, the winemakers are then willing to share their new ideas and techniques. The winery has benefited from the intellectual capital and has some intellectual property as a result in the application of the new idea/ technique. The winemaker has enhanced their own intellectual capital which they will share within their own networks. In an industry such as this, where the love of wine transcends competitive business practice, tacit knowledge is openly shared and disseminated. This advantages the industry. It also means that tacit knowledge can become codified knowledge relatively quickly.

TACIT AND CODIFIED KNOWLEDGE

The idea management process begins with individual inspiration and the tacit knowledge of the individual. Tacit knowledge has been defined as non-codified, intangible know-how that is acquired through the informal adoption of learned behaviour and procedures (Wheelright and Clark 1995). Polanyi (1962) describes tacit knowing as involving two kinds of awareness: the focal and subsidiary. While individuals may be focused on a particular object or process, they also possess a subsidiary awareness that is subliminal and marginal (Howells 1996). A discovery that involves focused awareness is usually termed synchronicity since the individual is actively seeking an idea or a solution to a problem (Ayan 1997). Tacit knowing also involves subception, that is, learning without awareness and this is associated with serendipity. Serendipity is defined as a random coincidence or accident that triggers an idea or concept when the individual is not actively seeking it i.e. without awareness of a problem or need (Ayan 1997).

Once the idea is generated, usually through a dynamic moment or illumination and is recorded, it becomes explicit or codified knowledge (Nonaka and Takeuchi 1994). Codified knowledge is information that can be written down and expressed clearly, while tacit knowledge is more complex – it comes from a combination of the application of codified knowledge and an individual's prior experience and creativity (Mascitelli 2000). The ability of the organization to assimilate and exploit this knowledge is termed the firm's absorptive capacity (Cohen and Levinthal 1990). Prior knowledge allows organizations to better recognize the value of ideas generated, facilitating purposeful adoption and exploitation of these ideas through the firm's internal capabilities and processes. Innovation diffusion and adoption require the transfer of associated tacit knowledge or technical 'know how', which is by nature a difficult task (Mascitelli 2000; Powell 1987).

Biotechnology and pharmaceutical firms obtain their competitive advantage from their codified knowledge base. The written nature of codified knowledge means that it can be defined and subsequently licensed or sold as an intellectual asset. However tacit knowledge or intellectual capital is essential for developing a sustainable stream of codified knowledge. In the pharmaceutical industry, patents are the most commonly used means of defining codified knowledge. They are used to obtain a competitive advantage by excluding other companies from manufacturing their products. The pharmaceutical company need not necessarily develop the technology themselves. They may only have a licence to use the technology in a particular region. Hence codified intellectual assets can be considered as an asset of a firm despite the firm not owning the knowledge or the people that generated it.

The following recruitment statements from Genentech's and Nexia's web sites give some indication of the corporate culture they claim to have and the focus on their people and their people's tacit knowledge.

Genentech

Since 1976, something truly unique has been evolving in the San Francisco Bay Area. It's a company with a different approach. One where people are the purpose for, and the strength behind, its existence. It's a company called Genentech – with a culture and a history that could only be called legendary. At the very beginning we planned to bring the best characteristics of an academic environment to the Genentech corporate culture. We hired the most competent, enthusiastic people we could find, providing a stimulating, challenging, yet supportive environment, and encouraged publications, peer interaction, and open communication. That tradition of looking for the best people, the right people, and encouraging their creativity has been what makes Genentech 'work'. I take great pride in their dedication and enthusiasm and their significant contribution to science and human health. (Genentech 2004)

Nexia

A common opinion is that an organizational memory can be viewed as an interrelated network of a group's decisions, rationales, processes, best practices, policies, and procedures that exists independently of the individuals who contribute to it. From this perspective, a group is defined as two or more individuals who have merged their thoughts, feelings, and actions to achieve a common goal. (Nexia 2004)

ABSORPTIVE CAPACITY

Absorptive capacity (ACAP) refers to the integration of external knowledge into a firm. Cohen and Levinthal (1990) describe this concept as an organization's ability to recognize and exploit external information for its own use. Prior knowledge allows organizations to better recognize the value of the information of an innovation. Once recognized, innovation adoption is more purposeful and thus faster. The adoption process is also assisted by an organization's 'boundary spanners' or 'gatekeepers' who facilitate the transfer of more complex information between internal and external environments (Drury and Farhoomand 1999). This role can be fulfilled by managers and scientists in the biotechnology company as they meet for industry breakfasts, local, national and international conferences and seminars, scientific meetings, initial collaborations and correspondence via e-mail and chat sites, or just work with university academics and their students.

Nonaka (1994) has stressed the importance of internal organizational structure for engendering a creative environment in organizations. Hansen (1999) has shown that ties between organizational subunits – particularly weak as well as dynamic ties, are important also in the creative process. This has been reiterated elegantly by Hargadon and Sutton (2000) in their recommendations for building an innovation factory – where ideas must be retained and circulated within an organization.

Polanyi focused on creativity at an individual and psychological level, whereas the aforementioned works focused on organizational structures. Poincaré (1908) and much later, Koestler (1964) stressed the importance of associative thinking to create new and unexpected associations and 'creative sparks' of new ideas. Nonaka and Takeuchi (1994) argue that while knowledge is developed by individuals, organizations play a critical role in amplifying and articulating that knowledge. Once an idea is generated, usually through a dynamic moment and is recorded, it becomes explicit or codified knowledge (Nonaka and Takeuchi 1994). The ability of the organization to assimilate and exploit this knowledge is termed the firm's absorptive capacity (Cohen and Levinthal 1990).

There are a number of factors that affect absorptive capacity and adoption. A firm's prior related knowledge enables it to recognize valuable new information, assimilate it and apply it to commercial ends; therefore a firm with a better developed knowledge base in a particular field will have a higher absorptive capacity for new opportunities. Absorptive capacity allows firms to pursue projects with a higher probability of success due to their superior knowledge (Deeds 2001). Scientists at Merck were able to take advantage of ground-breaking research on the process of cholesterol formation in the mid-1970s, due to prior work conducted in isolating mevalonic acid, a link in the cholesterol chain, in 1956. By 1975 Merck came up with the product Mevacor (Gambardella 1992).

The example of Merck highlights another issue where firm competitive advantage is concerned. Absorptive capacity involves learning and acting on the scientific opportunities occurring outside the firm, and using the gathered information to redirect technology development activities within it (Deeds 2001).

Effective knowledge management is heavily dependent on the ability to collaborate, both inside and outside the organization (Miles et al. 2000). In the biotechnology industry a major source of external information tends to be the scientific community – a social network of scientists and researchers from universities and non-profit research institutions with shared norms and values (McMillan and Hamilton 2000).

ACAP doesn't just apply to individual firms, as exemplified by the network RPR-Gencell. An integrated division of 18 public and private partners specializing in various aspects of cell and gene therapy, RPR-Gencell was federated by pharmaceutical company RPR. Scientists and academics within RPR played a key role in coordination between partners, using their credibility within the academic community to push for cell and gene therapy research, representing a basic element in a strategy based on a strong relational, non-contractual component. At RPR-Gencell, due to the highly specific and complementary assets involved in transactions, interdependencies are reinforced through specific relationships and technological synergies; members gain access to each other's complementary technologies in a much broader way than if each member had developed several bilateral alliances by themselves (Staropoli 1998).

Experience with technology adoption and decision-making processes required to adopt are also related to absorptive capacity and innovativeness (Lefebvre et al. 1991). This is where concentration on a narrowly defined niche has created an advantage for some biotech companies.

Prior knowledge and experience in adoption is enhanced by an organization's research and development investments, which help maintain absorptive capabilities (Cohen and Levinthal 1990). Research and development provides incentives to learn and accumulate knowledge and these incentives increase by quantity and clarity of information about an innovation (Cohen and Levinthal 1990). This clearly creates an advantage for established biotechnology companies over start-ups. As in other areas such as management board and scientific advisory board formation, it is the track record that the individuals bring to the firm which are vital for countering the lack of track record the NBF has as an entity. It is the objective of many NBFs to build this track record through employing and keeping good scientists and technicians, to bring on or bring in quality professional managers and to stack the SAB with high quality, committed star scientists if at all possible.

Nonaka and Takeuchi (1994) have suggested a new 'hypertext' organization as a structural base for knowledge creation in which the key requirement is the capability to acquire, create, exploit and accumulate new knowledge on a continual and cyclical process. Kapeleris et al. (2004) have taken this one step further and suggest that the organization's internal cycling process is causally connected to the capacity of an organization to absorb ideas from external sources. They have hypothesized and developed a dynamic model (see Figure 5.3) that relates the *rate* of internal cycling or *processing* of ideas (a function of internal organizational structure) to the external absorptive capacity (α) of an organization. If ρ is the *rate* of processing of ideas within a company then $\alpha = f(\rho)$ and the derivative $\delta\alpha/\delta\rho$ is positive. The rate of processing could be seen as creating a demand or appetite for new ideas from external sources or from individual employees. This relation links the individual (psychological and internal) features of creativity with the organizational structures necessary for capturing knowledge and developing new ideas. The barrier between the individual agents (be they external or internal to the organization) can be regarded as a 'semi-permeable membrane' with α describing the rate of diffusion across this membrane, α being a multi-dimensional variable.

Absorptive capacity depends on several factors. Prior knowledge in both related and unrelated fields allows faster recognition and assimilation of information (Mascitelli 2000). This is dependent on two related ideas. Learning is cumulative and learning performance is greatest when the object of learning is related to what is already known. Therefore, a diverse platform of knowledge is a huge advantage for the further acquisition of relevant information. For NBFs the emergence from a research project or programme creates a focused knowledge base which can be rightfully considered to be expertise, despite the fact that sophisticated equipment such as NMR, PCR, matrix-assisted laser desorption ionization mass spectrometry (MALDI-MS), PAGE, molecular imaging, and various other forms of high through-put screening have de-emphasized scientific skills and have focused more on technical skills in the operation of the equipment. Despite

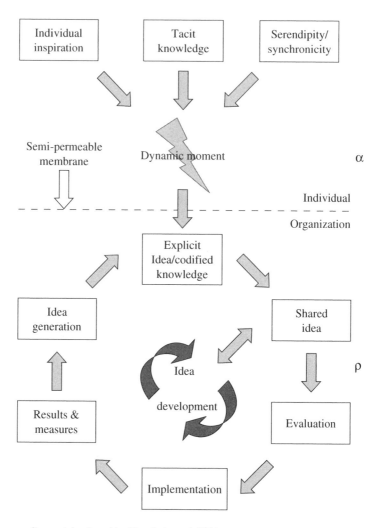

Source: Concept developed by Kapeleris et al. 2004

Figure 5.3 The semi-permeable membrane and ACAP

this, the reliance on post-doctoral and post-graduate researchers, as well as the vast majority of NBFs in most countries being actively involved in research projects with institutions, provides a clear indicator that the science is still vitally important in the innovation process.

KNOWLEDGE TRANSFER

As tacit knowledge becomes increasingly codified, tacitness decreases and knowledge transfer is easier (Zucker et al. 2002).

Discovering scientists become important in technology transfer when a new discovery has both high commercial value and a combination of scarcity and tacitness that defines natural excludability. 'Natural excludability' is the degree to which there is a barrier to the flow of the valuable knowledge from the discoverers to other scientists. Tacit, complex knowledge provides partial natural protection of information, both separately and jointly with more formal property rights (Zucker et al. 2002).

A robust indicator of a firm's tacit knowledge capture (and strong predictor of its success) is the number of research articles written jointly by firm scientists and those discovering, 'star' scientists, nearly all working at top universities (Zucker et al. 2002).

Furthermore, Zucker et al. point out that information tends to be produced in a tacit form, increasing in tacitness as a function of distance from prior knowledge and that it requires resources to codify. They also suggest that tacit knowledge often requires that one of those already holding the knowledge works with the novices to teach them in a hands-on process. Codification can sometimes be impossible. For example, patented cell-lines are often stored in a depository for public inspection as codification may not be possible to precisely explain how the lines are created (Zucker et al. 2002).

Tacit and codified knowledge are also transferred by different methods, and these methods are seen to influence different stages of the adoption process (Rogers 1995). For example, impersonal mass communications methods, which often transfer codified knowledge, are more effective at the awareness stage, whereas inter-personal communication, necessary for tacit knowledge transfer, is more important at evaluation and decision-making for adoption. This demonstrates that tacit knowledge is more difficult to capture, transfer and clearly understand because of the necessity of face-to-face interaction (Mascitelli 2000).

A firm's ability to exploit external knowledge is often generated as a byproduct of its R&D (Cohen and Levinthal 1990). An organization's research and development investments provide incentives to learn and accumulate knowledge and these incentives increase by quantity and clarity of information about an innovation.

IP is the description of an output, a product. Creativity is a process, an approach, and a mindset. It is said to be the seed of innovation. What is the relationship between IP and the process of innovation?

If knowledge is tacit it cannot be recognized as IP, as this implies a form of codification – when does tacit knowledge become known as IP? The ability to share tacit knowledge relates strongly to inter-firm and internal diffusion – the ability of organizations and individuals to engage in the processes of recognizing, selecting, receiving/giving, embedding and applying new knowledge. Organizations themselves cannot diffuse knowledge, however they can perhaps resource the activity, create the opportunity; build the relevant skill base to enable the diffusion process.

For extensive internal diffusion to occur requires an emphasis beyond product and even process innovation, beyond the technical. Organizational and administrative innovations, marketing innovations, shared problem solving, elimination of functional boundaries, integrative design manufacture and marketing all have to be achieved.

Internal diffusion also relates to the organizational structure and design. For example it may be easier to achieve in a company which is essentially an R&D unit than in one which has a more complex structure and functional focus. This relates to the type and main functions or purpose of the firm. Firm structure and design would relate to factors such as control and management as well as internal systems. Effective project management is an effective part of this process. This topic was covered briefly in Chapter 3.

The internal diffusion can then be extended to inter-firm diffusion where the organizational boundaries are blurred in strong networks and alliances where tacit knowledge is shared. This can extend any innovation diffusion model developed into the network literature. It would then incorporate trust, commitment, mutual benefit, embeddedness, strong and weak ties, social networks, intangible knowledge based assets.

THE ROLE OF IT IN KNOWLEDGE SHARING

Smaller organizations with less than 50 employees are small enough to permit effective exchange of knowledge without IT assistance. However larger firms need a means of effectively communicating experiential problems and knowledge across various departments. The critical role of knowledge management in sustained competitive advantage is clear from the expense that numerous companies incur to develop, catalogue, and 'own' information.

The systematization of knowledge management may involve creating directories of staff who may be contacted on various issues or highlighting each staff member's experiences outside typical daily operations. To avoid instances of reinventing the wheel, these directories can describe particular tasks that have been performed in the firm's recent history. Alternatively, if possible, this information could be made explicit in the form of frequently-asked questions or best practice guides accessible on a firm's intranet. This allows contributors to update the material on a regular basis allowing for a snapshot of competencies to be identified at any point in time. Updating is also important as technological change can allow for tacit knowledge to be applied in a new context.

More complicated knowledge management systems will use a firm's IT infrastructure to integrate its e-commerce, customer relationship, knowledge and supply chain processes with its operational value chain processes. This way the information obtained from purchasing and deliveries, and so on, can all be processed and used by all staff to make business decisions.

LEARNING TO SHARE KNOWLEDGE

How can internal diffusion assist the organization to become more competitive? Changing the focus of learning from an individual basis to a strategic basis (not simply upgrading individual skills, but embedding the skills development in the firm by aligning with strategy, by contextualizing to the firm's needs and by applying directly to company specific projects) will enhance the internal diffusion. The embedding process also creates the uniqueness which can create intellectual property. Although it may not be easily converted to unique, tacit knowledge. Maybe the goal should be more to convert from tacit knowledge to intellectual property as quickly as possible as the ideas can be embedded in the firm, but to avoid making them generic enough to be codified knowledge.

This creates a different imperative; to embed knowledge in the firm sufficiently that it becomes an intangible asset for the firm. It doesn't actually diminish the intellectual capital of the individual as this relates to the capacity to generate future ideas, while the intellectual property refers to existing ideas. See Figure 5.4.

So the role of training can be to convert from codified knowledge which provides no advantage to the firm, to intellectual property which does provide advantage to the firm. It also does not dilute the intellectual capital that resides in the individual. The training must be contextualized to the firm's specific needs and be applied to the firm's existing projects within the training programme itself, not afterwards. The contextualization can also enhance internal diffusion as it is applied to the firm's tasks directly and does not require conversion from generic to specific skills. If the firm is viewed as a 'distributed system of knowledge' (Tsoukas 1996) then firms can only exploit individual knowledge once it is transformed into organizational knowledge.

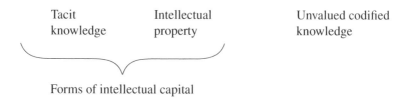

Figure 5.4   Forms of intellectual property

Firms must not only manage internal knowledge successfully but also be able to use external information to their advantage. This ability to exploit information across the organization is referred to as the organization's absorptive capacity (Cohen and Levinthal 1990). Absorptive capacity refers not only to the ability of a firm to recognize external information but to assimilate and then to apply it to commercial ends.

REFERENCES

Ayan, J. (1997), *Aha!: 10 Ways to Free Your Creative Spirit and Find your Great Ideas*, New York: Three Rivers Press.

Barney, J. (1991), 'Firm resources and sustained competitive advantage', *Journal of Management*, **17** (1), 99–120.

Bontis, N. (1998), 'Intellectual capital: an exploratory study that develops measures and models', *Management Decision*, **36** (2), 63–76.

Bontis, N., N.C. Dragonetti, K. Jacobsen and G. Roos (1999), 'The knowledge toolbox: A review of the tools available to measure and manage intangible resources', *European Management Journal*, **17** (4), 391–402.

Bontis, N., W. Chua, S. Richardson (2000), 'Intellectual capital and the nature of business in Malaysia', *Journal of Intellectual Capital*, **3** (3), 223–47.

Carayannis, E. and J. Alexander (1999), 'Secrets of success and failure in commercialising US government R&D laboratory technologies: a structured case study approach', *International Journal of Technology Management*, **18**, 326–52.

Carroll, R.F. and R.R. Tansey (2000), 'Intellectual capital in the new internet economy – its meaning, measurement and management for enhancing quality', *Journal of Intellectual Capital*, **1** (4), 296–312.

Cohen, W. and D. Levinthal (1990), 'Absorptive capacity: a new perspective on learning and innovation', *Administrative Science Quarterly*, **35** (1), 128–52.

Deeds, D. (2001), 'The role of R&D intensity, technical development and absorptive capacity in creating entrepreneurial wealth in high technology start-ups', *Journal of Engineering and Technology Management*, **18**, 29–47.

Drury, D. and A. Farhoomand (1999), 'Information technology push/pull reactions', *Journal of Systems and Software*, **47** (1), 3–10.

Gambardella, A. (1992), 'Competitive advantages from in-house scientific research: the US pharmaceutical industry in the 1980s', *Research Policy*, **21** (5), 391–407.

Genentech (2004), Available: www.genentech.com, Accessed: 12 November 2004.

Hansen, M. (1999) 'The search-transfer problem: the role of weak ties in sharing knowledge across organization subunits', *Administrative Science Quarterly*, **44** (1), 82–111.

Hargadon, A. and R. Sutton (2000), 'Building an innovative factory', *Harvard Business Review*, **78** (3), 157–68.

Hine, D., and R. King (2001), 'Barriers and enablers to innovation: I the Queensland Wine industry', unpublished work, Queensland University of Technology.

Howells, J. (1996), 'Tacit knowledge, innovation and technology transfer', *Technology Analysis and Strategic Management*, **8** (2), 91–106.

Kapleris, J., D. Hine and R. Barnard (2004), 'Defining the biotechnology value chain: Cases from small to medium Australia biotechnology companies', *International Journal of Globalisation and Small Business*, **1** (1), 79–91.

Koestler, A. (1964), *The Act of Creation*, New York: Macmillan.

Lefebvre, L., J. Harvey and E. Lefebvre (1991), 'Technological experience and the technology adoption decisions in small manufacturing firms', *R&D Management*, **21** (3), 241–50.

Martin, M. and L. McIntyre (1996), *Readings in the Philosophy of Social Science*, Cambridge, MA: MIT Press.

Mascitelli, R. (2000), *Journal of Product Innovation Management*, **17**, 179–93.

McMillan, G. and R. Hamilton (2000), 'Using bibliometrics to measure firm knowledge: An analysis of the US pharmaceutical industry', *Technology Analysis and Strategic Management*, **12** (4), 465–75.

Miles, R., C. Snow and G. Miles (2000), 'The future.org', *Long Range Planning*, **33** (3), 300–21.

Nexia (2004), Available: www.nexia.com, Accessed: 12 November 2004.

Nonaka, I. and H. Takeuchi (1994), *The Knowledge-creating Company: How Japanese Companies Create the Dynamics of Innovation*, New York: Oxford University Press.

Palacios-Marques, D. and F.J. Garrigós-Simón (2003), 'Validating and measuring IC in the biotechnology and telecommunication industries', *Journal of Intellectual Capital*, **4** (3), 332.

Poincaré, H. (1908), *Science and Method*, London: Nelson.

Polanyi, M. (1962), *Personal Knowledge*, London: Routledge.

Polanyi, M. (1966), *The Tacit Dimension*, Garden City, NY: Doubleday.

Powell, W. (1987), 'Hybrid organizational arrangements: new form or transitions', *California Management Review*, **30** (1), 67–87.

Rogers E. (1995), *Diffusion of Innovations*, 4th edn, New York: Free Press.

Roos, J., G. Roos, N. Dragonetti and L. Edvinsson (1997), *Intellectual Capital: Navigating the New Business Landscape*, London: Macmillan Press.

Staropoli, C. (1998), 'Cooperation in R& D in the pharmaceutical industry – the network as an organisational innovation governing technological innovation', *Technovation*, **18** (1), 13–23.

Stewart, T. (1997), *Intellectual Capital: the New Wealth of Organizations*, New York: Currency, Doubleday.

Tsoukas, H. (1996), 'The firm as a distributed knowledge system: a constructionist approach', *Strategic Management Journal*, **17**, Winter, 11–25.

Wheelwright, S.C. and K.B. Clark (1995), *Leading Product Development: The Senior Manager's Guide to Creating and Shaping the Enterprise*, New York: Free Press.

Zucker, L., M. Darby and J. Armstrong (2002), 'Commercializing knowledge: university science, knowledge capture, and firm performance in biotechnology', *Management Science*, **48** (1), 138–53.

Zucker, L., M. Darby and M. Brewer (1998), 'Intellectual human capital and the birth of US biotechnology enterprises', *American Economic Review*, **88** (1), 290–306.

6. Intellectual assets II – intellectual gravity and managing IP in biotechnology firms

Co-authored by Maher Khaled, Cambridge Enterprise, Cambridge University

INTRODUCTION

Recent studies of commercially successful high technology firms have identified the critical role of knowledge acquisition in intellectual capital development and competitive advantage (Deeds 2001; Zahra and George 2002). In the biotechnology industry, firms develop their intellectual capital through knowledge exchanges via networks and collaborations with other firms as well as academic and research institutions. These associations are founded upon the relationships between key individuals within these organizations. Baum et al. (2000) have identified a strong correlation between the number of collaborations for a start-up biotechnology firm and their commercial success. The presence of well-reputed individuals or teams within or associated with a biotechnology firm will not only facilitate the formation of these linkages, but *draw* knowledge and resources to a firm through a process we have termed 'intellectual gravity'. We continue to examine how intellectual gravity, in conjunction with a firm's absorptive capacity, develops a firm's intellectual capital.

The model of intellectual gravity complements some recent amendments to the theory of absorptive capacity of high-technology firms in their attempts to capture maximum knowledge. Intellectual gravity provides opportunities for firms to engage in knowledge acquisition. It therefore provides *potential* for firms to acquire intellectual capital. In this sense, an opportunity for knowledge acquisition by a firm is an 'activation trigger' as described by (Zahra and George 2002). The firm's internal processes, described in detail in Zahra and George's article, are responsible for determining the efficacy of the diffusion of that knowledge throughout the firm. We emphasize two points that are key elements of intellectual gravity.

1. There must be a demand for the acquisition of the knowledge: The firm must have an appetite for it to be absorbed through its external 'semi-permeable membrane' (Kapeleris et al. 2004). Oliver's study of the number of collaborations partaken by biotechnology firms over the course of their lifecycle shows that their appetites change over time. The number of alliances a firm forms peaks at about year four as contacts are developed and remains raised until year eight as they search for appropriate activation triggers. It then drops off to a low at year 11 to peak again at year 14 (Oliver 2001). Once a product is released the company then returns to seek knowledge from external sources to inspire second-generation products (Nahapiet and Goshal 1998). These collaborations are often short-lived and rarely produce more than one publication (Liebeskind et al. 1996). This can be seen as a process of experimentation. Only when promising discoveries are made in the initial work is the project continued to further develop the technology (Zahra and George 2002). This leads us to the second element.

2. The redundancy of knowledge referred to by Zahra and other writers on absorptive capacity is actually attributed to the diversity of a firm's combined intellectual gravity. These authors are still correct in stating that some of this knowledge must be redundant to ensure an array of ideas is presented to the firm that has not already circulated amongst its competitors. The variety must lie within the stars or linchpins associated with the firm who attract various forms of knowledge to it. The knowledge is then attracted through the process of intellectual gravity that has been described through this chapter. Redundancy is not an issue until the knowledge gets internalized into the firm. Therefore the absorptive capacity literature is supportable, as redundancy is an internal function, while the diversity of knowledge is a function of intellectual gravity.

REPUTATION, STAR SCIENTISTS AND THE MATTHEW EFFECT

A number of works examine those individuals who display an extraordinary scientific output. Zucker and Darby (1996), term these individuals as 'star scientists' and have identified only 327 'stars' worldwide who have recorded more than 40 genetic-sequence discoveries or authored at least 20 articles reporting such discoveries by 1990. During the 1990s, sequence discovery had become routinized and was no longer such a useful measure of research success (Zucker et al. 1998).

These elite star scientists have an exceptional scientific output, whilst only consisting of 0.8 per cent of all GenBank scientists they accounted for

17.3 per cent of all articles listed in GenBank in 1990 (Zucker and Darby 1996) (GenBank is the US National Institute of Health's genetic sequence database, a collection of all publicly available DNA sequences). An article authored by one or more affiliated stars roughly doubles its citation rate (Zucker and Darby 1996). Star scientists demonstrate a number of differing characteristics to ordinary scientists beyond higher levels of publication. They often hold more patents (Zuckerman 1967) mentor fewer and brighter students, and are cited more regularly by other authors (Zucker and Darby 1996).

Merton's studies of Nobel Prize winners found that after these scientists were awarded their laureate status, their level of eminence did not fall much below that irrespective of the continuing quality of their work (Merton 1968). He argued that 'once a Nobel laureate, always a Nobel laureate' meant that co-authors were accredited with a disproportionately lower contribution to the findings. It was assumed that the star scientist was responsible for the majority of the work, if not for the source of the ideas, despite little evidence of any such suggestion. The subsequent result was that the coauthors received disproportionately lower credit for their contributions which led Merton to label this phenomenon the 'Matthew Effect' after the author of a similar biblical verse (Merton 1968).

The Matthew Effect has been affirmed by studies that indicate that the findings of eminent scientists are held to be more significant than that of lower status scientists despite the findings being the same (Foschi 1991). An example provided by Merton involves a paper that was originally rejected by a scientific journal. Once it was discovered that the editors had received a transcript that was accidentally missing Lord Rayleigh's name, the decision to reject the paper was reversed without its contents being altered (Merton 1968).

One of the premises of intellectual gravity is that the described Matthew Effect extends beyond the attribution of accolades. Although we support the observations made by Merton and others, as the Matthew Effect was first described during a period of limited academic entrepreneurship, it does not take into account the commercial influences on the reputation of academic scientists. Therefore it does not accurately describe the attractive forces and the manner in which reputation is accredited in twenty-first century biotechnology. Today, as scientists become more productive, their eminence grows and more colleagues seek to collaborate with them, increasing the sourcing of ideas. Furthermore, greater eminence through academic and commercial success attracts more funding, allowing for increased publication output and subsequently more recognition (Oliver 2004). These effects reinforce each other and snowball, resulting in a 'compounded' Matthew Effect (Van Looy et al. 2004).

Scientific Entrepreneurs and their Impact on Firms

The star scientists and laureates discussed are rarely entrepreneurs. Only 3 per cent of the stars identified worked in firms, the rest were located in academic or other not-for-profit institutions (Zucker et al. 1998). This does not mean that they were not involved in entrepreneurial activities as these rarely necessitate that they give up their academic tenures. To access a star's experience and know-how through their scientific and commercial networks, firms are willing to associate themselves with star scientists under favourable agreements (Murray 2004).

When a firm has a tie to a star scientist they will reach IPO (initial public offering) earlier and attain more funds from the float (Darby and Zucker 2002). When a star-linked firm goes public it has been shown that for each article written by a star, as or with a firm employee, additional funds of an average of $1.1 million are obtained (Darby and Zucker 2002). This is because financial investors base their decisions on the involvement of high-reputation scientists who invest their intellectual capital into the firm (Catherine et al. 2004). The investment of a star's intellectual capital dramatically increases a firm's measured innovative output (Zucker et al. 2002). In measures of innovation, stars appear to be dramatically more productive than their non-star peers from even the most respected universities.

Star–firm associations are also highly beneficial for the scientist involved. When a US star is affiliated with both a firm and a patented discovery, they are cited over nine times as frequently as their pure academic peers who lack patents or commercial linkages (Zucker and Darby 1996). These ties are highly influential to the founding of firms, the number of stars and collaborators active through 1980 provides a strong indication of where the biotech enterprises were distributed in 1990 (Zucker et al. 1998). Furthermore, by 1985, nine of the ten largest (according to market cap) biotech firms had articles coauthored between stars and firms (Zucker and Darby 1998).

NETWORKS, COLLABORATIONS AND SOCIAL CAPITAL

Recent approaches to knowledge creation involve a number of actors situated in a network (Van Looy et al. 2004). Firms, particularly start-ups, engage in strategic alliances to enhance their capabilities as well as other's perceptions of their capabilities. Alliances provide both access to resources and exposure, two key factors critical to the sustained success of start-ups (Baum et al. 2000). Resource-poor start-ups can access to state-of-the-art tax-payer funded research from universities, reducing their sunk R&D costs (Liebeskind et al. 1996). The inability to establish these relationships have been shown to be a precurser or a catalyst of organizational death (Oliver 2001).

Start-ups often play an intermediary role between academic and/or research institutions and larger firms with production capabilities but lacking in technological know-how (Liebeskind et al. 1996). When an inventor involves themselves in a new firm they contribute in two ways:

1. Human capital through technical know-how as well as scientific advisory board and executive guidance.
2. Social capital by providing a network of key individuals from whom the firm can establish specific relationships to acquire human capital (Murray 2004).

Although the inventor may remain external to the firm by retaining their academic tenure, their involvement can attract other persons, knowledge and resources to the firm. Liebeskind et al. (1996) describes these interactions as a 'social network' or a collectivity of individuals among whom exchanges take place that are supported only by shared norms of trustworthy behaviour.

Murray (2004) has identified two forms of an academic inventor's network:

1. Local laboratory network – current and former students and advisors established throughout a career.
2. Cosmopolitan network – colleagues and coauthors established through the social patterns of collaboration, collegiality and competition that exemplify scientific careers.

The personal associations developed throughout a scientist's career determine the extent of these networks and subsequently how entrepreneurial firms become embedded in the scientific community (Murray 2004). With the first advances in recombinant DNA in the 1970s, star bioscientists were very protective of the technique and their ideas. They tended to collaborate more within their own institution, which limited knowledge exchange (Zucker and Darby 1996). As the skills became more common, the knowledge more easily travelled through networks and diffused throughout the community.

Particular characteristics have been identified through a number of studies that typify well-networked individuals. Although a number of motivations exist for collaboration including the desire to learn tacit knowledge about a technique; to pool knowledge for tackling large and complex problems and to enhance productivity (Bozeman and Corley 2004), strangely it is the already more successful, larger laboratories that have more total, academic, international and industrial collaborations (Oliver 2004) pointing to a Matthew Effect in networks.

A strong indicator of the number of collaborations a lab will be involved in is the number of postdoctoral students working in it. Oliver (2004) found that the correlation, postdoctoral students to collaborations, far outweighs that of PhD and Masters students.

These postdoctoral students are also most likely to be attracted to a scientist who has three or more patents. Postdoctoral students coming from other labs will have acquired skills which are of value, allowing them a greater choice of lab supervisors (Oliver 2004). This again reinforces the proposition that certain scientists of high repute and intellectual capital will attract additional people and knowledge to their labs.

Biotechnology firms tend to cluster together to facilitate networks and knowledge exchange. On average, co-authorship decreases exponentially with the distance separating pairs of institutional partners (Katz 1993, cited in Bozeman and Corley 2004). Geography can provide a strong indicator of network activity and firm success (Cooke 2003). Close knit communities allow for informal communication which is the main source of collaborative agreements (Bozeman et al. 2004). Researchers tend to collaborate with scientists located nearby and are rarely involved in distant or cosmopolitan collaborations. Those who are however, tend to be research group leaders who tend to have larger grants (Bozeman et al. 2004).

The Dynamics of Intellectual Gravity

The literature cited above examines several patterns of knowledge exchange in the biotechnology industry. We propose a model to explain the drawing power exhibited by individuals or research teams, most evident in the behaviour of star scientists. Murray (2004) was of the opinion that the manner through which scientists contribute social capital, rather than human capital to entrepreneurial firms, was poorly understood. Although intellectual gravity is a distinct construct from social capital, we believe that it explains the process in which entrepreneurial firms acquire intellectual capital. See Figure 6.1.

As intellectual gravity resides within the individual and not the firm, a gravitational pull is exerted from wherever that person presides. This can be either internally, as an employee or externally, as an associate of the firm. Associates may be involved in roles with other organizations, in these instances the degree of their drawing power that is contributing to the firm

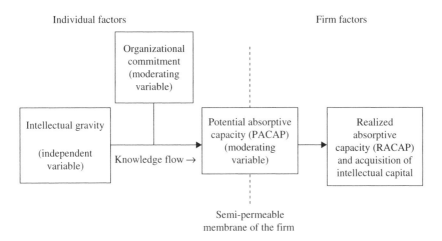

Source: Hine and Khaled (2004)

Figure 6.1   A model for intellectual gravity

is determined by their personal commitment. An individual's gravitational pull reaches outside the organization to attract knowledge to the firm, fuelling its potential absorptive capacity (PACAP) (Zahra and George 2002). The knowledge is absorbed by the firm in a manner that is akin to a 'semi-permeable membrane' (Kapeleris et al. 2004) which suggests that there must first be a demand for the knowledge.

Several characteristics determine an individual's drawing power or 'intellectual mass'. As each individual's strengths vary between these factors, so will the fields in which that person will exert a stronger gravitational effect. A simple example is that a basic research scientist is likely to attract other researchers and students to their lab, whilst a successful entrepreneurial scientist will attract more commercially-orientated people and knowledge.

Students
More eminent scientists will have more students wanting to work with them. The ability to pick and choose sees prestigious labs having generally less but more talented and experienced students. Merton (1968) believes that stars can play a 'charismatic' role in science and excite intellectual enthusiasm among those they work with. Zuckerman (1967) found that of 55 American laureates, 34 had previously worked in some capacity, under a total of 46 Nobel Prize winners.

Publications
The higher a scientist's publication rate, the more recognition they tend to receive in the scientific community. Publishing work that builds on the established literature encourages the exchange of ideas which can lead to collaboration. Some firms make a strategic decision to withhold publication of their scientist's work. This not only diminishes the firm's status but it prevents their researchers participating in social networks that attract new knowledge to the firm (Deeds 2001). Academic scientists are involved in, on average, twice as many collaborations at any point in time than industrial scientists (Zucker and Darby 1996). This suggests they would have a greater capacity to attract knowledge through their networks.

Patents
A study of Israeli bioscientists found that they had an average tenure of 20 years however they had only been listed as an inventor on an average of one patent each (Oliver 2004). Those who do have patents also have significantly more international and industrial collaborations (Oliver 2004). In this sense, acquiring a patent over a technology acts as a catalyst for other events that enhance one's 'intellectual mass'.

Colleagues
Merely working with another person of high repute can enhance one's reputation through association. In some instances it can be detrimental, co-authoring with a star can result in disproportionate recognition under the Matthew Effect but only in instances where the co-authors are clearly of differing levels of status. Stars have also been observed to attract other stars to their firms. Once a firm has acquired a star scientist, the likelihood that another star begins working with the firm is dramatically increased (Zucker et al. 2002).

Executive, scientific advisory board and political appointments
Scientists who involve themselves in areas beyond the research community will be able to access a vast array of resources from government and investment bodies. This can be in the form of funding, additional network connections or commercialization know-how.

This list does not attempt to be exhaustive, other avenues of extending one's networks and reputation will apply to enhance one's 'intellectual mass' to differing degrees. Some factors that have not been discussed above include membership of professional bodies, consultancies, membership of other firms, access to grants, keynote speaking at conferences and levels of education.

All of these activities enhance a firm's potential for knowledge acquisition through networks, collaborations, the hiring of new staff and so on. Some firms may take a proactive approach to harness a star's intellectual gravity by hiring one of their former students. Not only will this student be skilled in the technique and act as an asset of intellectual capital, they can also provide a link between the firm and the lab in which the star is situated (Liebeskind et al. 1996).

Social capital, is not easily traded between individuals (Nahapiet and Ghoshal 1998) and similarly we propose that intellectual mass rests with the individual or research team and is nontransferable. Social capital however, is jointly owned between the two parties in the relationship. Other parties can be involved to contribute to the value of that relationship. Firms can merely harness or 'borrow' one's gravitational affect to attract knowledge to the firm. Once this opportunity has arisen, separate firm-specific characteristics determine the utility of the knowledge. Critical opportunities, identifiable only in hindsight, are necessary to attain knowledge that is essential for firm survival. This places a responsibility on entrepreneurs to acquire a number of key individuals or 'linchpins' who are skilled or are able to access skills in key areas to ensure successful technological commercialization (Large et al. 2000).

Bozeman (2004, p. 12) accurately describes the key distinction between someone of only high intellectual capital and what we have defined as someone of high intellectual mass, 'In most fields, a brilliant scientist who cannot recruit, work with, or communicate with colleagues or who cannot *attract* [emphasis added] resources or manage them once obtained, is not a heroic figure but a tenure casualty or one or another variety of underachiever'.

In our model this can be adjusted to read as 'an individual or research team who, despite possessing significant knowledge assets and the associated reputation, cannot attract additional knowledge, resources or people that can be developed into additional intellectual capital for a firm, is of limited intellectual mass'.

COMMITMENT

An extensive body of literature on organizational commitment exists. However, the vast majority of it focuses on employees in established corporations. With regard to the commercialization of biotechnology inventions, involvement is almost always purely voluntary. Concepts of 'continuance commitment', where employees are committed due to a lack of alternative employment (Meyer and Allen 1991) rarely, if at all, apply.

Commitment acts as a moderating variable for intellectual gravity as many inventors are not willing to leave their tenured posts for a full-time position in a high-risk entrepreneurial firm. They will often be required to devote their time and energy to other organizations and projects. High organizational commitment demands two factors:

1. a strong belief in the acceptance of the goals and values of the organization for which the employee works.
2. a strong desire to maintain membership in the organization (Porter and Smith 1970).

An important element with relevance to entrepreneurial commitment is the level of personal or emotional involvement. Affective organizational commitment develops in response to work experiences that produce personal fulfilment. Therefore commitment increases with the length of tenure with the organization as well as in the occupation (Beck and Wilson 2001). Personal satisfaction is one of a number of factors that can be deemed of value when obtained from the investment of energy into a venture. Only if some form of value is being exchanged from the venture can the relationship remain sustainable (Nahapiet and Goshal 1998). As a result, an academic inventor may not be as emotionally attached to a venture or project if they are not familiar with the management style and practices of an entrepreneurial enterprise.

Studies in organizational commitment have found that there is no significant differences between the level of commitment of part-time and full-time workers amongst women in the retail industry (Still 1983). Murray (2004) asks whether full-time involvement on the behalf of an academic inventor, is more or less preferable to a more distant relationship but one in which the inventor maintains and continues to build his social capital through scientific networks. The scientist may only continue to provide value to the firm if they retain a strong academic affiliation and reputation. This goes against other findings that suggest that a scientist critical to the success of the venture must be totally committed to the successful transfer of the technology to the firm (Large et al. 2000). This is based upon the finding that those ventures in which the perceived commitment of the inventing scientist was high, enjoyed a significantly higher success rate.

In this instance we believe that an individual exerting a gravitational effect is best served to have a high degree of commitment to the firm to enhance knowledge flow. We acknowledge that detracting time and energy from activities that have allowed them to accumulate their intellectual mass might affect their contribution in the future.

During the earliest stages of start-up/spin-off, biotech firms will often be managed by academic scientists. As commercial influences may be limited at this point in time the firm must focus on developing the technology to the point where investment will be desirable. At this time it is the scientific networks of the inventor/academic scientists that are the key in establishing the scientific networks relevant to acquiring this knowledge.

KNOWLEDGE MANAGEMENT

Intellectual capital and intellectual gravity both contribute to the knowledge base of the NBF, that is, the amount and quality of knowledge available to the firm. Just what the firm does with this knowledge is largely encapsulated in two areas of interest:

1. The absorptive capacity as discussed in the previous chapter, which explores the ability of the firm to access and utilize knowledge.
2. Knowledge management is the other area which looks at creating an environment conducive to knowledge creation (tacit knowledge), knowledge sharing (both tacit and codified knowledge) and knowledge control and utilization (codified knowledge).

Once the knowledge becomes codified as intellectual property, then a new set of issues arises in its creation, management, protection and storage. This is covered in the latter part of this chapter.

In an organizational context, the effective discovery and transfer of information requires management of a collective of individuals' tacit knowledge (Pérez-Bustamante 1999). Knowledge management is the process of critically managing knowledge to meet existing needs as well as exploiting resident knowledge to create new opportunities. As the identification of valuable information is becoming increasingly difficult, the ability to access knowledge is now more important than a person's ability to retain knowledge. It has now become a case of *who* you know, not *what* you know.

Knowledge management aims to increase creativity through access to new ideas, reducing decision making time, increasing the reuse of acquired techniques and minimizing knowledge atrophy. Knowledge that does not circulate throughout the firm will progressively become stale and will eventually be obsolete. Furthermore, simply presenting material to staff does not necessarily result in learning. People must become actively involved in developing solutions to problems for their behaviours to change.

A basic premise of knowledge management is that tacit knowledge can, in part, be made explicit (King 2001) This does contradict the view that tacit knowledge can only be exchanged through personal interactions, however, any form of codification assists in the awareness of its existence. For this knowledge to be considered as tacit, its full context could not be appreciated from the explicit material.

Organizations themselves cannot diffuse knowledge. However, they can resource the activity, create the opportunity and build the relevant skill base to enable the diffusion process. It is dependent upon each individual employee to make use of these facilities which is why it is important for management to encourage the activity if the expense and effort is incurred in establishing these systems. An organization can also use these systems to shape the direction of learning within the firm. By having a strategic focus on the sharing of knowledge a firm can develop competencies in line with the company's specific projects.

Internal diffusion within a firm requires innovation to extend beyond products and processes. Shared problem solving through integrated design and marketing as well as business model innovation is necessary to facilitate the sharing of knowledge. The downside to this technique is that as organizations become more complex through adaptive changes, the ability for knowledge to diffuse through a complicated system is reduced. Therefore it is necessary to maintain an emphasis on streamlining internal structures and creating a pooled knowledge management system.

To achieve superior performance, especially persistent superior performance, a firm often needs multiple competitive advantages. Beating rivals on multiple strategically important vectors is essential for a winning firm. In order to build a host of competitive advantages and achieve superior performance, a firm has to gain individual competitive advantages one at a time. Enhancing one's understanding of the generic bases of the various competitive advantages is expected to help in the endeavour of advantage building. It will focus primarily on gaining advantage in the business arena. More broadly speaking, however, the essence of competitive advantage in any human endeavours is indeed the differential along any comparable dimension from which people derive value or to which people attach value, real or perceived.

KNOWLEDGE FLOWS

Codified knowledge, such as experimental results can be exchanged through networks and in relationships of trust. These are forms of social capital that do not directly fall under the scope of intellectual gravity. The distinction in this instance, although minor, is that gravity will provide a firm with *opportunities* to exchange knowledge with others whom they previously were not engaged in a relationship with. Codified knowledge is usually purchased or exchanged. We believe that intellectual gravity primarily facilitates the exchange of tacit knowledge between parties. This is more difficult to transfer than codified knowledge, particularly between organizations (Catherine 2004; Zucker et al. 2002a).

One technique is to simply purchase tacit knowledge and assist its diffusion through the firm. Senior management of the pharmaceutical companies who adopted biotechnology hired many new scientists who embodied the technology allowing existing personnel to acquire the expertise or leave (Liebeskind et al. 1996). Another method would be to merge or acquire entire firms developing a technology. Alternatively, firms can undertake collaborative research with external scientists. When performed successfully, knowledge is integrated directly into the ongoing R&D programmes of both organizations, something that cannot occur through pure market exchanges (Zucker and Darby 1997). The role of management in R&D intensive firms has gone from 'coordinating the on-going internal activities of the firm through a command and control structure' to 'providing appropriate organizational support for both the internal and external exchanges' that are essential to the firm's survival and success (Cohen and Levinthal 1990, p. 131). In either instance, the acquisition of tacit knowledge from outside the firm is reliant upon its absorptive capacity.

INTELLECTUAL PROPERTY

The term 'intellectual property' refers to the outputs of creative endeavour in literary, artistic, industrial, scientific and engineering fields, which can be identified and protected under legislation relating to patents, plant breeders' rights, trademarks, copyright, and design rights. Of the different forms of intellectual property, patents for new technology are of prime interest to scientists and engineers, although copyright can also be important (for example, for the protection of computer software). Like any other form of property, intellectual property can be sold, leased or mortgaged – so long as ownership has been established unambiguously. By providing security of knowledge, and establishing rights and rewards, intellectual property stimulates the innovation process. (BBSRC, 2004).

Patents provide some measure of the 'effectiveness' of a firm's R&D (Freeman 1982). Monitoring aggregate data over a range of US firms including pharmaceutical companies, Griliches et al. (1987) found a very strong relationship between R&D expenditures and patent applications, even after controlling for propensity to patent. Investment in R&D maintains the absorptive capacity of the firm (Cohen and Levinthal 1990), a critical aspect of intellectual capital.

Pharmaceutical R&D is a rather formalized and systematized process, where the route to market is narrowly prescribed by regulatory authorities. There is a high propensity to patent, and it is exceptional to find a pharmaceutical/biotech firm that does not attach importance to securing patient protection for its inventions.

Peptech

The Australian company Peptech Ltd provides a salutary example of the risks to be managed in association with patenting, and could lend weight to the observation that the value of a firm's patents depends on its ability to defend them. While the company holds patents in Australia, Canada, Europe, and the USA for the structure–activity relationships of TNF-alpha binding ligands, licensing the technology to major pharma companies Knoll (subsequently acquired by Abbott Laboratories) and Centocor (a subsidiary of Johnson & Johnson) for their respective rheumatoid arthritis treatments Humira and Remicade, proved problematic. While Centocor signed a licensing agreement in January 2001, and paid $30 million in royalties for sales in Europe, Canada, and Australia, the company subsequently decided that its product did not infringe Peptech's patents, and ceased to make royalty payments. The dispute went to arbitration in September 2003, and was resolved in October 2004 to Peptech's advantage. In the case of Abbott, a satisfactory outcome was obtained for Peptech: the licensing agreement was also signed in January 2001, and Abbott's product Humira (adalimumab) was launched in the USA in January 2003. A milestone payment was made in May 2002 when a regulatory submission was made in Europe, but, when potentially lucrative European regulatory approval was won in September 2003, Abbott issued a statement that it did not believe its product infringed Peptech's patents. Litigation was successful, and resulted in a large undisclosed sum being paid by Abbott in consideration of Peptech's patents. This outcome contributed greatly to Peptech's current situation of holding $40 million in cash reserves, which ensures that the company will be well-placed to defend future challenges to its intellectual property.

Intellectual Property Management

Knowledge management is a broader concept than that of IP management, yet IP management is an essential step to creating a strong IP strategy upon which the NBF's competitiveness can be based. The co-ordination of the

IP assets of the firm, while not a creative influence on the NBF, provides a solid basis for the commercialization effort. IP management is both an internal and an outsourced function in NBFs. Many start-ups will utilize the same law firms and patent attorneys they outsource their patent submission to. Few have internal legal staff who have the knowledge base to plan and establish an IP portfolio or a more complex strategy. The cost of outsourcing is prohibitive for many NBFs, hence few are able to establish effective IP strategies, just as few NBFs are able to create a convincing business plan, tending to create instead a product development plan.

While the management of intellectual property cannot increase the amount of invention disclosures made by a firm, this being dependent upon having talented and innovative scientific staff, effective IP management can however increase the ratio of revenue producing patents from those inventions that are made. The merits of the invention will ultimately determine its value. However, poor management of its protection and a failure to leverage the technology through commercial dealings, will result in a failure of the technology to achieve its potential returns.

Alchemia

Another Australian biotech company Alchemia Pty Ltd provides an example of the importance of intellectual capital to the growing biotech company. Alchemia's core technologies are focused around the manufacture of novel crystalline carbohydrates. In 1998 Alchemia brought to Australia two chemists from the University of London, where they had worked for Alchemia for three years, and possessed the know-how to develop large-scale preparation of carbohydrates. The company has also been able to attract talented young researchers who have seized the opportunity to visit and work in Australia, from Britain, Sweden, France, Germany, and Asia. The core technologies, for which immediate applications were in preventing superinfection from prolonged use of antibiotics, and preventing transplant graft rejection – niche markets worth AU$50 million annually – have now spawned further opportunities in drug development and construction of carbohydrate libraries. Relational capital is also important, and collaboration with the University of Queensland's Centre for Drug Design and Development has greatly strengthened Alchemia's intellectual capital base. Consequently the company has been able to attract the funding critical to its survival and growth, not only from Federal Government grants, but also from venture capitalists, the Australian Technology Group, and Medica Holdings Ltd. Alchemia also provides an illustration of a useful conceptual tool in the construct of 'intellectual gravity'. This is a holistic concept that encompasses core competencies, and describes the accumulated 'mass' of intellectual resources that draws human and financial capital to a company. A firm's intellectual gravity interacts with both intellectual capital and the intellectual property that arises out of it, in a continuing 'virtuous cycle'.

This practice of designing mechanisms to improve the generation, intra-firm diffusion and inter-firm protection of knowledge is the subject of a new field of theory and practice called 'knowledge management' (Carayannis and Alexander 1999). The need for knowledge management arises from the failure of the accounting standards to fully recognize the value of intellectual capital.

As any one biological patent will often cite a number of other prior patents, work in any particular field will often require licensing deals to obtain access to the existing technology. To facilitate the progress of technology, early-stage patents are often licensed at nominal sums or alternatively as part of a cross-licensing deal. Patents that describe technology of proven commercial value are often licensed for significant value. In many cases just a sole patent or a family of patents describing a technology can provide the major basis of income for a biotechnology or biopharmaceutical company.

It can cost as much as US$1 billion to put a compound through clinical trials with a view to obtaining FDA approval. There are approximately six companies worldwide who can afford to attempt the full process on a regular basis therefore patent protection is necessary to permit other organizations an opportunity to take their inventions to market. Licensing deals are often exchanged on terms of a nominal royalty rate of a few per cent on the gross sales of a drug. More recently, middle-sized R&D firms are becoming more dynamic in their licensing negotiations and are able to carry the development of their technology closer to market. In some cases they can afford to take a compound through Phase 1 or even Phase II trials. Others have developed distribution networks or manufacturing facilities. This permits them to enter into licensing agreements that may only share the final stages of taking a drug to market thus allowing them to negotiate higher royalty rates.

In some instances, firms will be able to commercialize aspects of their technology but will still license the technology out for other reasons. The brand power held by Big Pharma is often motivation to leave the drug's market entry to a name that consumers will trust. Alternatively, a firm may only entrust a major pharmaceutical to commercialize a drug in certain countries. By working with a larger partner a firm can gain access to their expertise which will allow them to develop competencies in commercialization by attempting to manage the drug's release in minor territories. This will enable them to better facilitate the release of future compounds.

IP and Confidentiality – the Impact on Knowledge Sharing

Intellectual property as an intellectual asset is a line item on the NBF's financial statements. As such those undertaking a management role will be concerned to ensure that this asset improves in value rather than diminishes. It is more likely to be the case when a professional manager is brought into the company once it has grown to sufficient size, that control of IP will become more of an issue. For professional managers, disclosure of any information on the intellectual property assets of the company will impact on the company's performance and needs to be prevented. Hence the proliferation of confidentiality disclosure agreements (CDAs) and non-disclosure agreements (NDAs), in the industry. Some would see these as a necessity, others as a necessary evil and still others as an unnecessary evil. The differing perspectives are often based upon the role an individual plays within the company. For scientists, collaboration, knowledge sharing, conference attendance and idea generation are not only the source of their inspiration, but often the source of new IP. For managers each of these can be seen to be the source of IP leakage that needs to be controlled. Remember that Cohen and Boyer met at a conference and it could be argued that it was this meeting and the disclosure of their intellectual property to each other that spawned the biotechnology industry. Then again Genentech is now a US$6.5 billion company because it patented and protected the IP which emanated from Cohen and Boyer's work, established lucrative licensing deals with major players early in its development, and kept hold of its IP rather than disclosing it.

According to Chesbrough (2003), companies either operate in a 'closed innovation' or 'open innovation' paradigm or both. Companies' management of IP is dependent critically on which paradigm they are in. In the close innovation paradigm, a company manages IP to create and maintain its idea and prevent others from using them. Whereas in the open innovation paradigm, a company manages IP to leverage its own business and learn new ways to apply and integrate new technologies offerings from others' use of the company's ideas.

This is a grey area in terms of pros and cons arguments. For instance there is the pie perspective, that it is better to have a smaller slice of a bigger pie, than to have a large slice of a small pie. That is knowledge sharing and technology diffusion go hand-in-hand. Each are good for an industry as when ideas spread others incrementally innovate and find new applications for the existing knowledge. As a result the industry expands and many companies as well as their customers gain. On the other hand, knowledge sharing dilutes the intellectual asset of the company. This asset valuation is often the major if not singular source of valuation for the company. If this is lost, then so is the leverage for capital expansion of the company, its potential alliances and license deals. There is no obvious solution to this. On the one hand non-disclosure is good for the individual company, at least in the short to medium term, on the other hand disclosure is good for the industry, though individual companies can potentially forfeit their existence. It is a micro-versus-meso issue.

In effect, publishing study findings in scholarly journals is a way of taking the small slice of the big pie perspective. If not supported by a pre-existing patent the publication places the data in the public domain, making it available to all to utilize, providing the author with kudos, but little tangible benefit. The Human Genome Project (HGP) is an example of making findings available to all researchers for a greater good rather than controlling the IP through non-disclosure. Though in a case like the HGP, the findings were so important to the scientific community that it would be considered to be a public good. We won't delve too deeply into this quagmire, leaving the debate to those more adept in the field of ethics.

In a study of Queensland biotech companies (Frahm and Hine, unpublished) a number of case respondents spoke of the need to secure CDAs (confidentiality disclosure agreements) before communicating with external stakeholders. This then acts as a barrier to innovation, specifically knowledge transfer that naturally occurs in social networks and loose ties. In this study, the concern about confidentiality differed depending on the case concerned. For two companies, the fact that they were post patent meant there were little costs associated with open communication with external collaborators. This supports Ledwith's (2000) contention that external communication is a high-risk activity for small firms with few patents. Another company's solution was to be selective in who to communicate with and 'wrap them so tight in CDAs' that the exchange of information can then be quite free. Yet another company was more strategic in approach preferring to distribute small parcels of work to multiple collaborators so that no one external collaborator knew the whole picture. This was further supported with a fear campaign, and putting that company's staff in the external sites on a regular basis to act as 'policemen'. This may be because there was a general belief within the company that CDAs are not effective in controlling information. Another two companies were more controlling of staff internally, preferring to limit the amount of communication that went out of the firm.

Without disclosure there are no networks as there is no trust and no commitment displayed. Powell et al. (1996) suggest that a network serves as a locus of innovation because it provides timely access, knowledge and resources that are otherwise unavailable. However this can only occur if the network members do get access. The introduction of CDAs to protect intellectual property must have a limiting effect on innovation in terms of knowledge transfer. Therefore, it can be deduced that the innovation necessary for new product development (NPD) in small firms must occur within the firm. This means recruitment of a diverse staff, with experience in multiple organizations and in multiple projects is necessary. When the firms are large enough to have the expertise and resources to afford the patent process, or established trusting relationships with external collaborators, this dependence on internal resources is not as great.

Much of this debate remains a debate between collaboration and competition. We have moved well beyond this dichotomous debate to a realization of the need to collaborate to compete. The competitive advantage that can be gained can benefit an entire network of collaborating firms rather than the individual firm. Competitive advantage can be considered as the asymmetry or differential in any firm attribute or factor that allows one firm to serve the customers better than others and hence create better customer value and achieve superior performance. The higher a firm scores vis-à-vis its rivals, the greater its competitive advantage.

When intellectual capital is a source of sustainable competitive advantage, then the means by which that capital is protected is of paramount importance. Firms need to develop capabilities in allowing sufficient freedom for knowledge transfer to produce innovation, without risking leakage of intellectual capital. In an open system this is no simple task, and this area requires extensive investigation.

Just as exchange of knowledge between employees promotes the generation of new ideas, so does interaction between firms. The risk involved in this process is that employees may inappropriately or inadvertently share knowledge with external parties resulting in a loss of competitive advantage. Managers attempt to de-risk these interactions by ensuring that all exchanges are preceded by confidentiality disclosure agreements (CDAs). However, the entanglement of such exchanges in CDAs can act as a barrier to knowledge transfer. This is why innovation is most often promoted through naturally occurring informal social networks.

Intellectual Property Strategies

IP strategies extend the content and intent of IP management in seeking to align the IP, its generation, protection, and maintenance with organizational objectives. While it is strategic in its intent it does not include the co-ordination of intellectual capital let alone intellectual gravity. Therefore it remains focused upon the outcomes of each of these two processes.

Intellectual property protection is critical, though not sufficient. It is the end to which almost all NBFs aspire, however an end which most should arrive at relatively early in their lives, yet few actually achieve due usually to poor planning and resource deficiencies. Effective IP management means more than just protecting patents, trade secrets, recipes and trademarks. It also requires the ability to commercialize the IP, establish its efficacy, seek alliances and collaborations to take the IP and its value further, offer the rights to its development to others more capable, usually through licensing, and effectively monitor and enforce the NBF's intellectual property rights, or at least make provision for such enforcement. The cost of enforcement is high, if not prohibitive, for most biotechnology start-ups. Cash reserves are required for R&D and little can be spared for other non-income producing provisions. However, low cash reserves makes the NBF much more vulnerable to patent infringement and attempts to test the patent's strength through the courts. Large firms, especially Big Pharma, see this as a legitimate strategy. Regardless of the moral correctness of the strategy, it is a reality in the biotechnology industry.

Strategizing is about planning for a robust future. Making financial provision for future legal challenges is just one way to plan to overcome future challenges. Establishing a strong corporate framework from the beginning is another means to future success. Company structures, which safeguard assets from legal execution of patent infringement challenges, are rarely undertaken.

In a technology-intensive industry like biotechnology, patents are a crucial measure of success. However they are of little use if they can be challenged easily. Creating a strong IP position from the outset is another means to ensure a strong future. In many cases patents are written when the researchers are working in an institutional environment and working with public funding. Commercialization is not, however, the primary concern of the researchers, rather it is usually problem solving.

In fact all these issues are intertwined. Separating out assets through a complex, but strong corporate structure is essential; this prepares the NBF for future legal challenges. However it also means the financial provisions for these challenges can be minimized. Clear nomination of a company on the patent is also an important early step, as is looking after the inventors to ensure the IP value of the company is not diluted through departures and reduction in the firm's IP ownership. Getting quality professional help is also important, however it is also critical that the NBF builds its own IP knowledge base as quickly as possible as this will be invaluable in creating and maintaining an IP strategy as well as the IP management process generally.

While it is beyond the brief of this book to delve too deeply into IP law and different forms of IP strategy that are currently undertaken in the biotechnology and pharmaceutical industries, it is critical to be aware of the need to create a strategic position as early as possible for the NBF to establish its competitiveness and sustain this competitiveness into the future.

VALUATION TECHNIQUES FOR FIRMS AND IP

As intellectual property can be the sole basis from which many firms derive an income, the valuation of a firm for the purposes of acquisition is highly reliant on the rights it holds to IP. There are a number of techniques for valuing IP, as outlined in Chapter 4, so that it may be entered on a firm's balance sheet. Some techniques involve significant economic and techno-logical rationalization whilst others are often little more than a 'guessti-mate'. The three broad valuation approaches mentioned in Chapter 4 are described below.

Cost Approach

The cost approach assumes there is a correlation between the value of intellectual property and the expense incurred in its development. Although easy to calculate, it is only remotely accurate when applied to cal-culating the costs and time involved in basic research, creating designs or drafting documents. Inexpensive and simple inventions can produce spec-tacular results and when applied to blockbuster drugs this method is grossly inadequate.

Market Approach

This values the IP by comparing it to similar items that have been licensed or sold in the past. The market approach allows the marketplace to provide an indicator of value. However, markets can be volatile and technology shifts in value at a rapid rate. A further disadvantage of this technique is that IP is by its nature, unique, therefore identifying a similar transaction may be impossible. Nevertheless, a number of companies have established databases that collate assignment and licensing deals undertaken by phar-maceutical and biotechnology companies. Due to the typically confidential nature of this information and as a small change in a royalty rate can result in a loss or gain of millions of dollars down the track, access to these data-bases costs in the tens of thousands of dollars per year.

Income/Net Present Value Approach

These assessments can only be accurately performed by experts with expe-rience in both the life sciences and auditing fields. Although such audits may read for many pages before identifying a sum or a bracket for the IP's value (and hence these assessments are costly to produce), any economic or technological change can render the report as obsolete. The net present value approach involves assessing the potential market return for a technology based upon anticipated markets, minus the costs for its implementation. The significant distinction between this approach and the market approach is that this technique predicts future value, rather than looking at past transactions to assess the value of the IP. Despite the significant degree of 'crystal ball gazing' involved in this approach, it is nevertheless the most accurate technique for assessing the value of a patent that describes a technology, particularly one that is clearly leading its competitors.

When conducting due diligence for acquisition or mergers, all IP must be valued and entered as an asset on the balance sheet. This is why the market capitalization of high technology firms can appear to be over-inflated as the majority of their written down value is from intangible assets and not plant, equipment or cash reserves.

Rights issues should also be considered as the collaborative nature of scientific research often means that a number of scientists from several companies or institutions may have contributed to the development of a technology. These issues should be investigated well before commercial release so that appropriate agreements can be made. These outstanding issues can delay or even prevent the acquisition of companies or items of IP.

A general rule that applies to all three cost assessment techniques discussed above is that the value of a patent or item of intellectual property varies with time. The IP value life cycle can also be viewed in parallel with the high-technology product life cycle and the commensurate revenue streams, indicated in Figure 6.2.

So in effect, while few compounds reach the market, and the highest value of the IP is closest to market, the greatest costs are incurred well before market. The life cycle dilemmas facing NBFs are clearly a complex hurdle which when combined are often insurmountable. So why do so many try? Because of the potential for financial gain and to make a difference in the lives of those who will benefit from the final product.

IP STRATEGIES

The IP strategy is a particularly important aspect of the overall strategy for a biotechnology venture. There are a number of alternative paths that can be selected by the new biotech venture as it is seeking to leverage its people and its IP. The IP decision is intrinsically caught up with other important areas of the new venture such as the structure it is going to take, the profile of the management team, resource limitations and requirements, market access, and the product profile the venture wishes to establish.

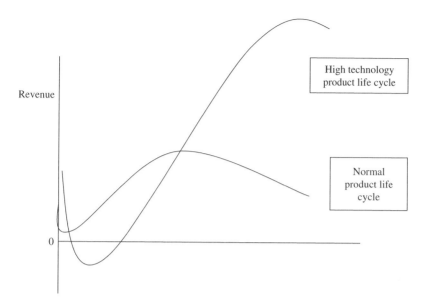

Figure 6.2   Revenues through high-technology PLCs

Invention to Licence

Once an invention has been confirmed, there are four distinct stages until that invention can produce revenue for the firm.

1. Invention is documented with associated lab books and data
2. Patent application submitted
3. Patent is granted
4. Revenue is acquired through licensing/commercialization of that patent.

The attrition rate of inventions through these stages is quite high. A study in the USA found that on average, approximately ten invention disclosures will result in one revenue producing patent (Morgan et al. 2001). This ratio will vary between industries and individual firms and institutions. The Lawrence Livermore National Laboratory showed how effective IP management improved their disclosure to licensing ratio from 50 to 1 in 1990 to 7 to 1 in 1997 (Jaffe and Lerner 2001).

    The average attrition rate arrived at by Morgan et al. (2001), was: of 100 invention records, 49 patent applications were lodged, from which

Lead candidates making it to the next stage of approval

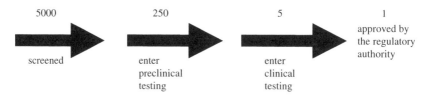

| 5000 | 250 | 5 | 1 |
| screened | enter preclinical testing | enter clinical testing | approved by the regulatory authority |

Figure 6.3   Compound success rate by stage

31 patents were granted, which produced 13 revenue producing patents (and hence 18 non-revenue producing patents). This does not even tell the entire commercialization story, as it is not until market launch, (and even later in cases such as Vioxx) that the true costs and revenues from commercialization can be gauged. Nevertheless Figure 6.3 provides a depressingly clear indication of the success rate from initial discovery to market approval.

The attrition rate provides a good reason why it is important to get the IP strategy right early in the R&D process. A correct valuation will lead to appropriate funding of the R&D and of the overall venture.

Proactive and Defensive Strategies

While this probably comes more into the realm of strategic choice there are both defensive and proactive strategies which can be utilized in patent strategies (as well as work arounds). In this regard the work of Stuart Macdonald is both pertinent and current. It is entitled, 'When means become ends: considering the impact of patent strategy on innovation'. It is hard to pick out one excerpt from another in terms of quality and relevance but we have attempted that with the quote below:

> such crass devices as patent pooling have long since been replaced by the sophistication of patent stacking, blocking, clustering and bracketing, blitzkreiging, consolidation, blanketing and flooding, fencing and surrounding, by patent harvesting and ramping up, by portfolio and network arrangements, and other devices that tend to be directed less at facilitating innovation than with discouraging the innovation of others (Macdonald 2002, p. 12).

The implications of Macdonald's work are not surprising – in areas related to pharmaceuticals, whether a good or a bad thing, the need for business strategy and management skills will become increasingly important. This is a theme continued further in a paper by Gangaram Singh (2002, p. 7) from San Diego. He states, 'the biotechnology industry is undergoing a significant change from research and development to commercialization. It is proposed that this change is likely to result in the greater importance of skills associated with business administration. In addition, there is likely to be a shortfall in the capability of achieving business administration skills'.

The biotechnology firm Genetics Institute decides which version of a drug to develop not only based on clinical trials results, but also according to which iteration can command the strongest patent protection (Rivette and Kline 2000). Patents also become useful in competitive hedging, where companies license or acquire competitive technologies that threaten its core business, and in so doing gain more options should they take off (Hatfield et al. 2001). In order to better compete with Johnson & Johnson, new entrant in the stent business, Guidant Corporation, acquired Endo Vascular Technologies, which held an unused stent patent that had been issued two years before Johnson & Johnson's. By doing so Guidant Corporation gained a foothold in the US coronary stent market by selling $350 million worth of the devices in its first six months of business (Rivette and Kline 2000). Thus the strategic value of intellectual property in enhancing firm competitiveness is clear here.

Alternative IP Strategies – Commercializing Internal Processes

The manner in which a firm organizes itself and conducts its business practices can, in some instances be patentable. Training programmes and manufacturing processes can be applied to other firms, at a price. Competitive advantage can be obtained through these techniques, just as it is attained from superior products and technology. These techniques can allow firms to manufacture products at a lower cost or alternatively automate certain processes. The internal workings of these organizations are often closely guarded. However, when staff members change firms they cannot prevent the ideas being transferred also. Therefore IP protection is sought to prevent this information from spreading throughout the industry.

In a sense this is an example of harnessing a firm's know-how in its truest form as it seeks to simply exchange knowledge and experience for cash. When a firm has invested effort in discovering the best available method for performing a process, they may choose to profit from this discovery as other firms may be interested in saving the time of reinventing the wheel.

Firms often find that their know-how is best dispersed throughout the firm in tailor-made knowledge management software. This facilitates the distribution of standard operating procedures and best practices in large corporations. Knowledge management is examined in the following section however protection of these computer programs is discussed below.

There has been extensive research around this area, as it aligns with firm strategies and capabilities for competitiveness. Some of the literature is quite academic, though an understanding of it is valuable to understanding the value of IP and IP management.

As part of an overall IP and R&D strategy, the venture can consider a license/licensing strategy as one means to a return on investment. Beyond the decision of whether to license or not, there are different types of licences which can be used. According to the Bioentrepreneur Toolkit, different types of inventions lead to different licensing strategies:

Exclusive license: A contract granting a licensee the sole right to use certain intellectual property to the exclusion of the licensor and all others.
Sole license: A contract granting a licensee the right to use certain intellectual property to the exclusion of all others except the licensor.
Non-exclusive license: A license agreement that does not prohibit the licensor from licensing others in addition to the licensee.
(BioEntrepreneur 2004).

Making the right choice in terms of commercialization paths comes back largely to a financial assessment. No matter what IP the venture holds, it only becomes meaningful and valuable in the context of what is done with it. What is done with it will dictate the returns made available from it, to be potentially reinvested into the R&D program of the venture. Some licensing strategy evaluation decisions will include: determining the best licensing or alternative strategy according to the product or contract service; the relative strength and independence of the future product; assessing likely and best-option licensees; developing a favourable, though reasonable license agreement profile including milestones, equity positions, collaborations, R&D spending and staff sharing and of course royalty rates when the product does actually arrive on the market; and finally a clear articulation of second and third horizon products, that is, those products which are initiated and developed after the company's first product has progressed through its pipeline (Canadian Innovation Centre 2004).

REFERENCES

Baum, J., T. Calabrese and B. Silverman (2000), 'Don't go it alone: alliance network composition and start-ups' performance in Canadian biotechnology', *Strategic Management Journal*, **21**, 267–94.
BBSRC (2004), Available: http://www.bbsrc.ac.uk/business/ip/ip.html, Accessed: 24 November 2004.

Beck, K. and C. Wilson (2001), 'Have we studied, should we study, and can we study the development of commitment? Methodological issues and the developmental study of work-related commitment', *Human Resource Management Review*, **11**, 257–8.

BioEntrepreneur (2004), Available: http://www.nature.com/bioent/toolkit/ip/index.html, Accessed: 3 June 2005.

Bozeman, B. (2004), 'Editor's introduction: building and deploying scientific and technical human capital', *Research Policy*, **33**, 565–8.

Bozeman, B. and E. Corley (2004), 'Scientists' collaboration strategies: implications for scientific and technical human capital', *Research Policy*, **33**, 599–616.

Carayannis, E.G. and J. Alexander (1999), *International Journal of Technology Management*, **18**, 326–52.

Catherine, D. (2004), 'Turning scientific and technological human capital into economic capital: the experience of biotechnology start-ups in France', *Research Policy*, **33**, 631–42.

Chesbrough, H.W. (2003), *Open Innovation: The New Imperative for Creating and Profiting from Technology*, Boston, MA: Harvard Business School Press.

Cohen, W. and D. Levinthal (1990), 'Absorptive capacity: a new perspective on learning and innovation', *Administrative Science Quarterly*, **35** (1), 128–52.

Cooke, P. (2003), 'Editorial: The evolution of biotechnology in three continents: Schumpeterian or Penrosian?', *European Planning Studies*, **117**, 757–63.

Darby, M. and L. Zucker (2002), 'Going public when you can in biotechnology', NBER Working Paper No. 8954: 1–35.

Deeds, D. (2001), 'The role of R&D intensity, technical development and absorptive capacity in creating entrepreneurial wealth in high technology start-ups', *Journal of Engineering and Technology Management*, **17** (1), 29–47 .

Foschi, M. (1991), 'Gender and double standards for competence', in C. Ridgeway (ed), *Gender, Interaction and Inequality*, New York: Springer-Verlag, pp. 58–72.

Frahm, J. and D. Hine (unpublished), 'The role of communication in new product development: the case of Australian biotechnology firms', paper submitted to *International Journal of Biotechnology*.

Freeman, C. (1982), *The Economics of Industrial Innovation*, London: Pinter.

Gangaram Singh (2002), 'Skills requirements of the biotechnology industry: moving from research and development to commercialization', working paper, Department of Management, College of Business Administration, San Diego State University.

Griliches, Z., A. Pakes and B.H. Hall (1987), 'The value of patents as indicators of inventive activity', in P. Dasgupta and P. Stoneman (eds), *Economic Policy and Technological Performance*, Cambridge: Cambridge University Press.

Hatfield, D., L. Tegarden and A. Echols (2001), 'Facing the uncertain environment from technological discontinuities: Hedging as a technology strategy', *Journal of High Technology Management Research*, **12** (1), 63–75.

Hine, D. and Khaled, M. (2004), 'Extending the conceptual thinking on intellectual assets: the importance of intellectual gravity for biotechnology firms', paper presented at The Academy of Management Conference, Hawaii, August.

Jaffe and Lerner (2001), 'Reinventing public R&D: patent policy and the commercialization of national laboratory technologies', *RAND Journal of Economics*, **32** (1), 167–98.

Kapeleris, J., D. Hine and R. Barnard (2004), 'Defining the biotechnology value chain: cases from small to medium Australian biotechnology companies', *International Journal of Globalisation and Small Business*, **1** (1), 79–91.

Katz, J. (1993), 'Bibliometric assessment of intranational university-university collaboration', PhD thesis, Science Policy Research Unit, University of Sussex, Brighton.

King, W. (2001), 'Strategies for creating a learning organization', *Information Systems Management*, **18** (1), Winter, 12–24.

Large, D., K. Belinko and K. Kalligatsi (2000), 'Building successful technology commercialization teams: pilot empirical support for the theory of cascading commitment', *Journal of Technology Transfer*, **25**, 169–80.

Ledwith, A. (2000), 'Management of new product development in small electronics firms', *Journal of European Industrial Training*, **24** (2/3/4), 137–51.

Liebeskind, J., A. Oliver, L. Zucker and M. Brewer (1996), 'Social networks, learning, and flexibility: sourcing scientific knowledge in new biotechnology firms', *Organization Science*, **7** (4), 428–43.

Macdonald, S. (2002), 'When means become ends: considering the impact of patent strategy on innovation', paper presented to the Workshop on Competition in Property Rights and Information Markets, Centre for Competition and Consumer Policy, Australian National University, Canberra, August.

Merton, R. (1968), 'The Matthew effect in science', *Science*, **159** (3810), 56–83.

Meyer, J.P. and N.J. Allen (1991), 'A three-component conceptualisation of organizational commitment', *Human Resource Management Review*, **1**, 61–89.

Morgan, R., C. Kruytbosch and N. Kannankutty (2001), 'Patenting and invention activity of US scientists and engineers in the academic sector: comparisons with industry', *Journal of Technology Transfer*, **26**, 173–83.

Murray, F. (2004), 'The role of academic inventors in entrepreneurial firms: sharing the laboratory life', *Research Policy*, **33**, 643–59.

Nahapiet, J. and S. Ghoshal (1998), 'Social capital, intellectual capital, and the organizational advantage', *Academy of Management Review*, **23**, 242–66.

Nonaka, I. (1991), 'The knowledge-creating company', *Harvard Business Review*, **69** (6), 2–9.

Oliver, A. (2001), 'Strategic alliances and the learning life-cycle of biotechnology firms', *Organization Studies*, **22** (3), 467–89.

Oliver, A. (2004), 'Biotechnology entrepreneurial scientists and their collaborations', *Research Policy*, **33**, 583–97.

Pérez-Bustamante, G. (1999), 'Knowledge management in agile innovative organisations', *Journal of Knowledge Management*, **3** (1), 6–19.

Porter, L. and F. Smith (1970), 'The aetiology of organizational commitment', unpublished paper, Irvine, CA: University of California, Irvine.

Powell, W., K. Koput and L. Smith-Doerr (1996), 'Interorganizational collaboration and the locus of innovation: networks of learning in biotechnology', *Administrative Science Quarterly*, **41** (1), 116–45.

Rivette, K. and D. Kline (2000), 'Discovering new value in intellectual property', *Harvard Business Review*, Jan–Feb, 54–66.

Still, L. (1983), 'Part-time versus full-time salespeople: individual attributes, organizational commitment and work attitudes', *Journal of Retailing*, **592**, 55–79.

Van Looy, B. and M. Ranga, J. Callaert and K. Debackere (2004), 'Combining entrepreneurial and scientific performance in academia: towards a compounded and reciprocal Matthew-effect?', *Research Policy*, **33** (3), 425–41.

Zahra, S. and G. George (2002), 'Absorptive capacity: a review reconceptualisation and extension', *Academy of Management Review*, **27** (2), 185.

Zucker, L. and M. Darby (1996), 'Star scientists and institutional transformation: patterns of invention and innovation in the formation of the biotechnology industry', *Proceedings of the National Academies of Science*, USA, **93**, 12709–16.

Zucker, L. and M.Darby (1997), 'Present at the biotechnology revolution: transformation of technological identity for a large incumbent pharmaceutical firm', *Research Policy*, **26** (4/5), 429–46.

Zucker, L. and M. Darby (1998), 'Entrepreneurs, star scientists and biotechnology', *NBER Reporter*, 0276119X Fall: pp. 1–5.

Zucker, L., M. Darby and J. Armstrong (2002), Commercializing knowledge: university science, knowledge capture, and firm performance in biotechnology', *Management Science*, **48** (1), 138–53.

Zucker, L., M. Darby and M. Brewer (1998), 'Intellectual human capital and the birth of US biotechnology enterprises', *American Economic Review*, **881**, 290–306.

Zucker, L., M. Darby and M. Torero (2002), 'Labor mobility from academe to commerce', *Journal of Labor Economics*, **20** (3), 629–60.

Zuckerman, H. (1967), 'Nobel Laureates in science – patterns of productivity, collaboration, and authorship', *American Sociological Review*, **32** (3), 391–403.

7. The cycle game I – product life cycle, R&D cycle and organizational life cycle

INTRODUCTION

Biotechnology companies, in general, face a far more extended R&D pipeline than many other high-tech industries. In the IT industry, typically $2–3 million would be required to get software designed, developed and on to the market within a six to twelve-month period. However, in biotechnology, a typical drug takes 15–20 years and $200–900 million to bring to market (Champion 2001), with no guarantee of success. Investor criticism of biotechnology as a poor investment performer due to long lead times compared with IT, ignores the very slow technological diffusion of Shockley's solid-state physics breakthroughs that were made in the early 1950s. The resultant real improvements offered by the most prominent solid-state product innovation, computers, are only manifesting themselves now. The breakthroughs of Cohen and Boyer in recombinant DNA, seen to be the point of inflection for the biotechnology industry, only occurred in 1973. It is therefore unrealistic to expect massive changes in product development lead times within such a short period.

Business life cycle theories usually propose a common process, whereby an item upon which attention is focused is perceived to progress naturally through distinct phases over a period of time (its life). Classically, the phases commence with a slow birth, and proceed via rapid exponential growth, through an inflection point and consequent diminishing rates of growth, to a period of maturity and stability, then to ossify and inevitably move into decline or renaissance. Two of the most important life cycle theories in the study of management, the product life cycle (PLC) and the organization life cycle (OLC) will be discussed in this chapter.

THE DEVELOPMENT OF LIFE CYCLE CONCEPTS

Life cycles cannot be viewed independently of the entire business operation as there is considerable cause and effect at work. The diagram in Figure 7.1

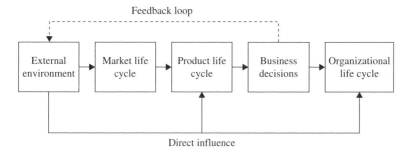

Figure 7.1   Influences on the organization

provides some indication of the influences of the various life cycles. In this diagram, there is only a dotted line for the feedback loop to show the influence the typical small biotech business has on its environment. This is because, though much has changed for small business, it is still rare to see them influence their external environment. Pioneers in markets – the entrepreneurial biotech companies – provide the exception to this rule as they can significantly influence a small evolving market. Therefore the market in which they operate is so small that they can exert market power. Of course this will only be maintained until the market grows, usually at a greater rate than the individual business, so that their market share diminishes over time.

Taking a broader view of the product we look at product cycles and the implications of the product life cycle analysis for the firm product positioning. The concept of the life cycle as applied to business has been evolving since Schumpeter, who based his ideas and analyses on earlier work by the Russian economist Kondtratieff on long-wave business cycles (Schumpeter 1939). This evolution has inevitably resulted in the establishment of new branches in the life cycle literature tree. The literature on life cycles from Emery and Trist (1965) to more recent observers such as Specht (1993) and Aldrich (1990), observes the organization as it moves through its various life stages. The needs of the organization differ at each stage, as do the appropriateness of management processes and practices.

THE PRODUCT LIFE CYCLE

With the product life cycle (PLC), the variable of interest is the total quantity of sales of a given product (class, form, brand). Normally, the PLC includes four stages, (1) a period of *introduction*, slow growth as a product is initially introduced, (2) followed by *growth* via rapid market acceptance, (3) then *maturity*, when sales growth slows as the product has achieved

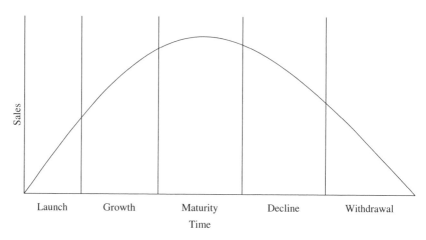

Figure 7.2 A typical product life cycle

acceptance by most potential buyers, and (4) finally *decline*, which sets in as the market becomes saturated, better substitutes emerge, and replacements are the only addition to sales. (Howard and Hine 1997). See Figure 7.2.

There are a number of general forces and effects associated with the PLC model, including:

- Diffusion of innovation takes place as the new product becomes increasingly recognized, legitimized and accepted by consumers/buyers over time (Rogers 1995);
- Experience curves come into play, production technologies improve and better practices emerge. Profitability increases, the product becomes cheaper, and more demand emerges (the demand function). Inevitably, product demand slows, and profits are maintained by controlling costs. Industry shakeout commences as competition for market share starts in earnest in a market that is stable or shrinking in size.
- The competitive behaviour of firms and markets evolve, through different market structures from monopolistic competition to oligopoly and monopoly.

To simplify, the PLC predicts that demand for a product will grow and then decline with market saturation. As product sales grow through the PLC, then so must productive capacity and the number and/or size of businesses that produce the product. Thus the PLC predicts a pattern in the way small and large businesses develop, which can then imply other economic potentials

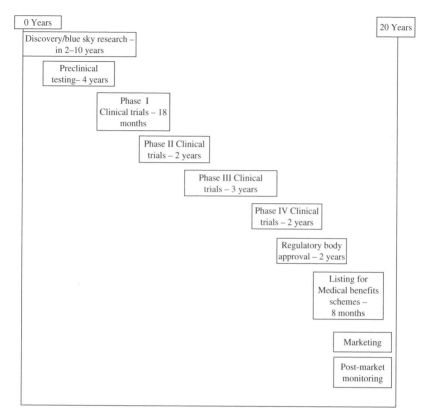

Figure 7.3   An example of the R&D Pipeline in biopharmaceuticals

such as in job creation. Logically, it could be expected that job creation would rise as product demand rises, whereas employment would fall as demand for product declines.

The product life cycle in biotechnology is aligned with the R&D pipeline. One reason for the dominance of the R&D pipeline concept over the more typical product life cycle is that in most areas of biotechnology there are no products which have actually reached the market as yet. Stages of development are better indicated through the R&D pipeline giving a clear indication through the stages of R&D, from discovery research, to applied research including patenting, then through pre-clinical (including toxicity testing usually involving pharmacokinetics) and the three to four clinical trials stages prior to regulatory approval and market launch. The entire R&D cycle, which can take 10 to 20 years, and cost $200 million to $1 billion, if successful (see Figure 7.3).

Looking at the diagram in Figure 7.2, the vertical axis is sales. Most small biotechnology firms are still some years away from having actual products. Their progress cannot be judged by a cycle of turnover of product, so the PLC is supplanted by the R&D pipeline. The major difference being that the ultimate end is not death, but market entry. The R&D pipeline also provides the small biotech firm with a basis for judging their exit strategy from the product as well as the overall product strategy, which is essentially the IP strategy – that is, when to license, whether to out-source trials, whether to manufacture, whether to remain a research-based company, and ultimately whether to allow the company to be acquired by a larger player who can move the product more effectively through the pipeline. The exit strategy decision was covered in more detail in Chapter 4. However for a founder of a small biotechnology firm, which has a research focus, it is always tempting to sell off the company with its IP and use the proceeds to fund further research in a related area, hence returning to one's comfort zone.

The outlook as given in Figure 6.3: after 10–20 year R&D cycle, with vast amounts of data compiled, particularly during the clinical trials, and between $200 million to $1 billion spent, we have a product. However for Merck, Biota and others, even once the pipeline seemed complete, the future wasn't rosy as the following case study demonstrates.

Biota

R&D cycle

R&D was a highly collaborative process. Relenza® (Zanamivir), a neuraminidase inhibitor (NI) binds to the neuraminidase enzyme on the surface of the influenza virus, inhibiting viral transfer from cell to cell. The drug was developed out of outstanding basic research between 1977 and 1985: (1) at ANU, on the influenza virus, and (2) at CSIRO where the molecular structure of neuraminidase was determined, and a part of the structure identified that was conserved in all strains of influenza, and was thus a potential target for rational drug design. Biota Holdings Ltd was established (and listed on the ASX) in 1985 to exploit this potential, and funded development at the Victorian College of Pharmacy, where the Zanamivir molecule was designed and synthesized. Activity against influenza A and B was first demonstrated in 1992, and clinical development was pursued with multinational Pharma GlaxoSmithKline (GSK), which has particular strengths in respiratory and antiviral therapeutics. The long (22-year) R&D cycle is typical of a biopharmaceutical firm. Between 1986 and 1992 Biota received R&D grants, and from 1992 to 1997 R&D tax concessions. The licensing agreement with GSK resulted in reimbursement of all Biota's

R&D costs from 1985 to 1990. The therapeutic potential of the Zanamavir molecule is now being explored in other indications, specifically avian influenza, initiating a new R&D/development cycle.

Product life cycle
Classically, an innovative product enters a new market where there is little or no competition, which does not appear until the maturity stage. However, Biota, approved by the US FDA on 26 July 1999 for influenza A and B and launched shortly thereafter (with GSK as manufacturing and marketing partner) in Sweden, the UK, the USA, and Australia, had immediate competition from Roche's NI Tamiflu®, approved by the FDA on 27 October 1999. The potential market for influenza is more than 500 million worldwide annually, and Relenza® was expected to produce substantial long-term cash flows. A high profile $42 million consumer ad campaign for Relenza® was conducted in winter 1999. However, market uptake was poor, largely reflecting: (1) The less familiar and less convenient inhaled method of administration of Relenza®, compared with Tamiflu®, which is administered as a capsule, and a liquid for children; (2) the more limited regulatory approval for Relenza®, while Tamiflu® was approved, not only for treatment, but also for prevention of influenza. Relenza® was approved for adults and children older than seven years, while Tamiflu was approved for treatment from the age of two years; (3) a serious side effect – bronchospasm and breathing problems necessitating immediate intervention – associated with Relenza®, but not Tamiflu®. In July 2000 the FDA sent an 'Important drug warning' to US doctors, upgrading the Precaution about bronchospasm in the original labelling to a Warning. Biota made an operating loss of $718 000 in the year ended 30 June 2001, compared with a profit of $2.5 million the previous year, the loss reflecting a 20 per cent fall in revenues from royalties and licensing fees for Relenza®. Relenza® sales revenue in the influenza season 1999–2000 was $20 million, compared with $41 million for Tamiflu®. Relenza® revenues have continued to fall. Because of the presence of Tamiflu® in the market, Relenza® was not able to develop a market niche, and develop full awareness of product benefits among potential consumers. Thus, Relenza® has not really moved beyond the introduction phase to the phase of rapid growth. Its side-effect profile may ultimately be a limiting factor.

THE CHALLENGE – IT'S GETTING HARDER

Some of the dilemmas facing the small biotech company as it decides whether to pursue the commercialization of its IP are listed below. This is a daunting list and helps to explain why founders of small biotech firms often license out their IP at a very early stage, even at pre-clinical trial stage, where the royalty returns are at their most minimal.

- High attrition rates of potential drugs;
- 15 years from idea to market;
- One in 20 drugs get through Phase I clinical trials; one in 50 get through Phase II; one in 5 get through Phase III trials; survival rate during development;
- Five years of market exclusivity;
- Only one out of four products shows a significant profit;
- Average cost to develop a drug is estimated at $897 million escalating investments and Wall Street expectations for growth;
- Increasing regulatory requirements and speed of FDA approval process;
- Expectations for sustained leadership and global presence are low, as two to three new drugs are launched each year, and the market growth rate for those drugs is only 15 per cent.

Decisions based upon the R&D pipeline, the biotech industry's effective product life cycle, will impact dramatically upon the organizational life cycle of the firm. Those who license out very early will tend to remain research-only companies. Those who choose to take the development further before licensing, but choose to outsource most of the ensuing tasks, will become largely a project management company. Those who choose to take the development further in house will shift from a research focus to a development, scale-up manufacturing focus. This will dramatically alter the business model as an entirely new equipment base, skills set, funding regime and network will be required to succeed. What these firms are doing is enveloping more of the value chain themselves, hence becoming larger more complex entities. The small biotech firm is not unique in this way, it is a path followed by many entrepreneurial firms. However it is an area few have succeeded in. In Canada for example, of over 300 biotech companies identified in the Deloitte Borderless Biotechnology report from 2002, Shire BioChem Pharma and QLT are the clear success stories. Deloitte also identified Aeterna, Biomera, Biovail and Stressgen as rising stars. The lack of obvious superstars is indicative of the newness of the industry as well as failure rates and bottlenecks in the pipeline. It is the medium to large biotech firms that have overcome their bottlenecks and have made the successful transition to a more complex focus.

The R&D cycle, starting at idea generation/inception leading through a number of stages prior to production is critical to product and firm success. The previously sequential nature of the R&D process, considered to be a chain-link approach to new product development has given way to more complex, chaotic approaches to the R&D cycle using innovation networks. The change has been important as a means to minimizing the cycle time and to increase the likelihood of bringing a potentially successful product to market. Ensuring a reduced R&D cycle and innovation speed must be coupled with the development of a management framework which also ensures an innovation and product stream (a series of products resulting from the innovative efforts of the biotech company and its partners). Speed and stream work together to assist the development of a number of products, each in a short period of time. Such a focus is based upon the need to achieve economies of scope in a environment of shortening product life cycles.

INNOVATION SPEED

Biotechnology companies in general face a far more extended R&D pipeline than many other high tech industries such as IT (Hine and Griffiths 2004). While the product life cycle may extend beyond that of other industries, the entire research and development cycle can reach 20–30 years. Given that patents are granted relatively early on in this cycle, companies must then achieve a dramatic return on investment over a time span as short as five years before competitor products can be legally launched on the market. Competitors, to access the codified knowledge required to replicate the product are in the meantime using patent documents and publications in scientific journals such as *Nature*, *Lancet*, and various other medical journals, to have their products market-ready for patent expiry.

While product life cycles are short (though not necessarily reducing), there are also limitations to the ability to reduce R&D cycles in biotech companies. Innovation speed (Kessler and Chakrabati 1996) a fundamental competitive factor in many industries is a significant challenge to biotech companies. Innovation speed is inextricably linked to diffusion. Based on the R&D cycle, innovation speed refers to the length of time it takes for a product to move from idea to commercialization.

Typical of the length of the biotech R&D cycle is the approval times of the major international drug approval bodies, such as the FDA in the USA, the TGA in Australia, and EUDC in Europe. Figure 3.1 indicated approval times for new molecular entities (NMEs). While it varies dramatically between countries, approval times are never less than 12 months. Further, this is only one step in the entire R&D process for most biotech products.

> The Food and Drug Administration approved 21 New Molecular Entities (NMEs) with active ingredients never before marketed in the United States. This number of NME approvals was up from the calendar year 2002 total of 17. Priority approvals – approvals for priority products of special medical importance – increased from 2002 as well: there were 14 priority NDAs and 9 priority NMEs, compared to 11 and 7 in 2002, respectively. The Agency's Center for Drug Evaluation and Research (CDER) and Center for Biologics Evaluation and Research (CBER) approved 466 new and generic drugs and biological products, many of which represent significant therapeutic advances. In particular, the Agency saw a significant increase in the number of approvals on NMEs which typically represent the most novel new drugs. (FDA website 2004)

In biotechnology, the set approval process, and the requirement for the trial phases prior to the approval process all serve to limit the extent to which biotech companies have control over the R&D cycle and are able to reduce it. It is only through reducing the R&D cycle, thereby enhancing innovation speed, that more time is provided for returns on investment before patent expiry.

Inextricably linked to diffusion is the concept of innovation speed. Based on the R&D cycle, innovation speed refers to the length of time it takes for a product to move from idea to commercialization.

It is not simply the type of innovation that is important in the relationship; of equal importance to competitiveness is the speed at which new ideas are brought to fruition as innovations (Kessler and Chakrabarti 1996). These authors also present an approach to ensuring this innovation speed is maintained or improved:

> Structure-related organizational-capability factors can facilitate or impede innovation speed in several ways, relating to the following conceptual categories:
> 1. The degree of empowerment or decision-making autonomy of the project team.
> 2. The degree of project integration.
> 3. The degree of organization in the development process. (FDA website 2004)

The combined speed and stream approach is also necessitated by cash flow requirements – the need for the individual firm or partners in an innovation network to achieve the positive cash flows from commercialized products that will sustain the growth of the firm/network and ensure the availability of funds to support the R&D effort which creates a negative cash flow. In this sense it should be recognized that R&D is only a means to an end, the end being the commercialization of innovations to the ultimate end of profitability & business growth.

In a more recent paper on innovation speed, Kessler and Bierly (2002), extend their work to include an integral assessment of the benefits of innovation speed to product quality and a reduction in investment costs, each leading to improved project success. However, this is moderated by the internal and external environmental uncertainty. Their findings attested to the generally positive impact of increased innovation speed, stating 'the hoopla surrounding speed is largely justified but needs qualification as forcing rapid development under high uncertainty may produce failure' (FDA website 2004).

So this reinforces the idea that innovation speed, and the reduction in R&D pipeline time, will be generally advantageous to the biotech firm. Given major constraints in reducing cycle times, particularly in the early research and regulatory approval times, the best opportunity to reduce R&D times is in the development and trials phases of the pipeline. Kessler and Bierly also indicate that improved speed leads to reduced investment costs. Clearly burn rates have a temporal basis, so reduced time should also reduce burn rates.

The goal is clear, but is the opportunity available to most biotech firms to reduce their cycle times? Methods used in engineering and construction to improve project management efficiencies such as the critical path method (CPM), briefly introduced in Chapter 3, are worth noting. However the most pertinent point made by Kessler and Bierly is the moderating effect uncertainty can have on project success. Unlike construction where most variables such as labour, equipment, materials, excavation rates, prefabrication rates, and so on (except of course the weather), can be calculated, biotechnology is notoriously uncertain.

Most research and product outcomes differ markedly from the original problem-solving intent of the researchers. Serendipity, particularly in early stage and applied research, is a fact of life. However in the applied research stage and its preclinical and then clinical trials, the testing, fermentation, purification, animal toxicity and patient testing and administration, and analysis of results, are stages of the pipeline which are methodical, requiring extensive quantitative quality controls and are much more equipment-based (while still requiring high-level skill sets) and are much more reminiscent of engineering and construction regimes.

Where coordination of tasks and stages is effective, product and process flows are controlled and resource efficiencies achieved, then there is scope to improve innovation speed while also enhancing product quality and reducing development costs. So while the entire pipeline is not open to improved speed, certainly the stages between applied research and regulatory approval are open to measures that can achieve process and non-technological innovations that can have positive temporal and financial impacts.

Figure 7.4 provides a succinct portrayal of the fundamental link between the R&D pipeline and the product life cycle in biotechnology. Intrinsic to this assessment is the average patent life, which creates the impetus for the

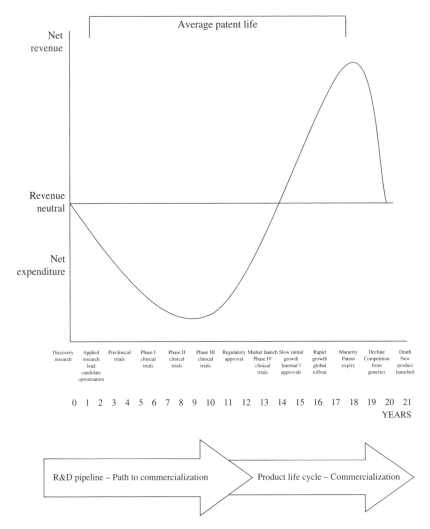

Figure 7.4   The relationship between the R&D pipeline and the product life cycle in biotechnology

R&D to be undertaken and creates strict limitations to the product life cycle. In Figure 7.4 it is also clear that the R&D cycle is proportionally much longer than the product life cycle, something unique to the biotechnology industry. The net revenue is short-lived, underscoring the reality that few drugs make more than the cost of their development, no matter how novel. Blockbuster drugs, those which make more than $1 billion revenue, are becoming increasingly rare. The inevitability of patent expiry, allowing the entry of competitors into the market, usually well before costs are recovered, signals the decline of the revenue base, as generics emerge as major competitors. The high margins, which existed while under patent protection, are lost. This is a stage in the PLC, where low cost production in a highly competitive market is the only means to survival.

Also highlighted in Figure 7.4 is the distinction between the path to commercialization, which is the focus of the R&D pipeline, and actual commercialization, post market launch, that denotes the PLC.

It should also be noted that the hypothetical product displayed in Figure 7.4 is one of the lucky ones. As we have indicated on a number of occasions, a very small proportion of lead candidates actually reach the market. Most fail or are abandoned somewhere in the R&D pipeline. This has major implications for the organizational life cycle, as the NBF has to sustain itself through the numerous failed attempts to bring on new products, while attempting to sustain its lead candidates through to market and beyond. This is another reason why we have talked about a likely shakeout in the industry. The implications this will potentially have on the industry life cycle will be discussed in more depth in Chapter 8.

THE ORGANIZATIONAL LIFE CYCLE

In developing the life cycle concepts there are two sequential paths taken:

1. The external factors which impact organizations are developed and explored;
2. The life cycle stages which are resultant of these external forces are proposed.

The common interest shared by the organizational life cycle (OLC) literature is the performance of firms over a length of time. The underlying assumption is that organizations move through various stages of growth in distinct, definable phases. Life cycle writers will typically examine contextual characteristics such as age, size, and growth looking for causal inferences. Although there may not be a common OLC, there are indeed common problems arising at similar stages of organizational development that encompass a large fraction of organizations (Churchill and Lewis 1983; Miller and Friesen 1984).

Helms and Renfrow (1994) argue that the purpose of the OLC is to provide insight into the development processes of the organization, thereby enabling the firm to cope with change and be proactive in planning and to capitalize on change. Moreover, the usefulness of a life cycle model is described by Greening et al. (1996) who found that the owners of small firms had no means to anticipate the needs of growing a small business and consequently could not plan for it. The growth or transformation of organizations are frequently conceptualized in the form of an OLC, comprising a number of phases or stages, associated with generic management problems and characteristics (Watts et al. 1998).

An old but still well respected perspective, Greiner's model (1972) represents a growth model of organizational evolution within the life cycle discipline and provides prescriptive statements for managers to plan and control growth. It deals with the individual behavioural aspects owner/ managers must face in the stages of growth. According to the model, initial growth is through creativity in the birth stage of an organization. The founders' technical or entrepreneurial orientation means communication is frequent and informal. As the firm grows, a crisis of leadership occurs as employee numbers increase. Hence, there is a need for efficient, large-scale production runs ahead of entrepreneurial drive and spirit. As a consequence, this first revolution will require subsequent changes to management style to achieve further growth. The next stage of growth though will be through direction, possibly under a new manager. A functional organization begins to develop through specialization with formal systems of communication and a hierarchy.

Greiner's work can be seen to be pioneering in the organizational life cycle literature. Greiner (1972) continued the focus on managerial style and competence in a five-phase model. In each phase, business evolves until a crisis point is reached when transition to the next phase is achieved by 'revolution'. The crises emerge as the firm grows more mature and larger. They are defined as a crisis of leadership, a crisis of autonomy, a crisis of control and a crisis of red tape. Pre-empting these crises are the evolutionary phases of growth through creativity, growth through direction, growth through delegation, growth through co-ordination and growth through collaboration.

Whilst Greiner's model gives insight into the decision process required to manage and achieve growth, the multitude of models available in the OLC discipline indicates Greiner's model is useful, but not conclusive.

Klofsten (1994), looks more broadly at the OLC and based upon a thorough review of the literature proposes eight aspects of early development:

1. Formulation and clarification of the business idea
2. Development of finished product
3. Definition of market
4. Development of an operational organization
5. Core group expertise

6. Commitment of the core group and prime mover of each actor
7. Customer relations
8. Other corporate relations.

The first four deal with the development process per se, five and six concern key corporate actors such as founders, CEO and board members or other members of the core group, the last two aspects concern the flow of external resources.

External Factors in the Organizational Life Cycle

When considering the OLC we cannot ignore the factors external to the organization which will assist or impede its growth and development prospects. This is particularly so in biotechnology given that the majority of companies are quite small and therefore have very little control over the environment in which they operate. We will go into more depth on the external environment in future chapters, however those external factors, which influence the OLC, will be explored briefly here. Some of the major findings in terms of the OLC are provided in the chronological list below:

- Churchill and Lewis (1983, pp. 40–2) provide a model which describes the stages as well as the factors 'which change in importance as the business grows and develops'. The model has five stages – existence, survival, success, take-off and resource maturity. The success stage has two sub-stages, disengagement and growth. There is no necessary precondition of growth-or-fail. They found eight general factors that change in importance through the life cycle, four of which 'relate to the enterprise, and four to the owner'. Enterprise factors include: financial resources, personnel resources, systems resources, business resources. Owner factors include: owner's goals, owner's operational abilities, owner's managerial ability, owner's strategic abilities.
- For Chell (1996), external environment for small enterprise consists of 'situations' which provide 'rules' that guide behaviour.
- Bird (1989, p. 138) also proposes general environmental factor categories (societal context variables) which include: '(1) economic, (2) political, and (3) technical "givens" of any moment of time in any location, (4) the Zeitgeist or spirit of the times, and (5) the cultural milieu'.
- Jeffress (1991, p. 13) describes entrepreneurship (the entrepreneur is defined as an initiator and agent of change) as a scarce resource that can be frustrated by social, economic and political factors. Love and Ashcroft (1999) similarly suggests a number of general categories of

factors, cultural, social, political, administrative, legal, and economic, that predispose an individual towards entrepreneurship, or if absent can reduce/eliminate the possibility of entry into entrepreneurship.

- Dodge and Robbins (1992, p. 33) confirmed 'that the owner-manager has to contend with different problems in the various stages of the organisation's life cycle. Seemingly, external environmental problems are more important early in the life cycle, with internal problems becoming more critical as the business grows and develops'. They propose four stages by which to categorize small enterprise owner-managers in the context of OLC theory: formation, early growth, later growth, and stability.
- Specht (1993) suggests general categories of factors, namely, social, economic, political, infrastructure development and market emergence, by which specific factors can be grouped. For example, the economic category includes capital availability, aggregate economic indicators, recession and unemployment.

Biota's organizational life cycle

Following a successful launch of the flagship product Relenza®, the company decided to focus on the viral respiratory disease area, and found partners to fund its non-core projects in diabetes, cancer, and Alzheimer's disease. If Biota had continued to commit itself to this single product class, its life span would have been equivalent to that of its products: Relenza® and second-generation NI FLUNET®. With the disappointing performance of Relenza®, Biota realized the need for a shift in strategic direction, broadening its niche to develop additional profitable product life cycles. Biota is now targeting, not only influenza and rhinovirus (the common cold), but also other viral agents such as HIV, hepatitis C, respiratory syncytial virus (RSV – for which it received a $2.7 million R&D start grant in 2003), and the SARS virus, as well as fungal, bacterial, and inflammatory diseases and cancer.

Biota is employing the classic strategy of using successive product life cycles to sustain its growth. Biota is moving from Phase I of Greiner's (1972) stages to Phase II, of 'Growth through direction'. After 'growth through creativity' (its outstanding collaborative R&D and flagship product launch; an evolutionary stage of growth), a 'crisis of leadership' (a revolutionary stage) occurred. Following the Relenza® debacle, and a threatened hostile takeover, there was a management shakeout, and restructure in early 2003, to expand business development internationally, and broaden the pipeline through external partnerships. The new CEO has experience in the North American biotech region that provides deal-making and technology collaboration strengths.

Biota established its US subsidiary Biota Inc in California in 2001, the key to its long-term international strategy. Business development and Biota's CEO are now based in the USA. Biota Inc has now established a new drug discovery platform, N-MAX, which has been licensed to GSK for $6.5 million to develop hepatitis C treatments. Australian R&D has been refocused on the multivalent coupling (MUCO) technology and chronic diseases. Biota has developed substantial intellectual capital drawing on its experience with Relenza®. Specific IP includes platform drug discovery technologies structure based drug design (SBDD) and MUCO, which are able to greatly reduce innovation cycle times, a factor critical to success in the biotechnology industry. Biota also owns the most sensitive diagnostic test for influenza, FLU OIA, which is very profitable, and now has a profit-share agreement around a second diagnostics kit, FLU OIA A/B. In addition, it has operational R&D facilities in Australia and the USA.

Industry life cycle for neuraminidase inhibitors
Roche's Tamiflu® continues to be the market leader for influenza A and B. Partner GSK having taken a strategic decision not to pursue development of influenza drugs, Biota has now signed a global cross-licensing agreement with Japanese Big Pharma Sankyo to develop and license both companies' long-acting NIs for influenza. On 20 February 2004, CSIRO researchers reported that Zanamivir (Relenza®) was effective against the H5N1 avian influenza strain – the first experimental evidence of Biota's claim that Relenza® would be effective against all influenza strains. It is uncertain whether these promising early findings will be sustained in clinical development; yet, with the news, Biota's share price increased more than 140 per cent in two days, 75 per cent of its stock changing hands. Biota will continue R&D to develop Zanamivir as a branded product for avian influenza, which has a mortality rate of 30–60 per cent, and is predicted to be worse than the 1918 influenza pandemic that claimed 40 million lives. This is an example where successive product life cycles, here involving the same generic compound but a different market brand, may potentially sustain the organization's growth.

PATHS TO GROWTH – SCALE OR SCOPE

A firm pursuing growth through expansion or product and distribution volume through scale production will eventually have to move from an organization based on the entrepreneurial tenets to a very different form of organization to manage the growth stage. Chaston (1997) supports Slevin and Covin's (1990) qualitative assertion that highly entrepreneurial/organic businesses, at certain points in the life cycle of the firm, might consider becoming less entrepreneurial and adopting a somewhat more mechanistic structure. The life cycle literature reveals a tentative consensus on the characteristics of the next stage.

A company that decides to grow through innovation by increasing the scope and diversification of products will still follow the organizational life cycle. However, it is proposed that this path will result in a different outcome in the organizational life cycle. The concept of maintaining innovation as a growth strategy has its merits. Chaston (1997) highlighted that successful small firms exhibit a proactive commitment to innovation as a means through which to sustain overall performance.

If we ask the question: are large pharmaceutical and biotechnology organizations redefining themselves to fill the space occupied by small companies in recent years? The answer, while complex, tends to indicate two distinct paths. Companies such as Genentech, Cellera, Millennium Pharmaceuticals and Ligand appear to be extending their reach through expanded R&D agendas, new partnerships, support for university research and teaching programmes and individual scientists. This denotes an economies of scope strategy, designed to achieve efficient growth through a broad product range. The Big Pharma appear to be undertaking a consolidation path, mergers such as Novartis and Aventis creating very large companies with sales forces in the 100 000s. In fact such mergers also have meant a dramatic reduction in staffing numbers to reduce duplication and the sell-off of a number of products deemed outside the purview of the new entity. Such a path denotes an economies of scale strategy, based on a narrowing of the product range and a concentration on core business capabilities, which for many of the long established Big Pharma is sales and marketing of products reaching the latter stages of their patent life.

Increasingly large organizations in this and other industries are utilizing the advantages of networks, alliances, joint ventures, research networks, social exchange and social networks and any form of collaboration which will enhance resource efficiency, improve possibilities for product and process innovation, extend product range, and improve market knowledge through knowledge sharing and mutually beneficial undertakings. These can be considered to be some of the features of economies of scope. Under the economies of scale regime, such undertakings were not required and the traditional principles of transaction cost economics, namely self-interest and internalizing exchanges as much as feasible, prevailed (Coase 1937; Williamson 1985). While many larger organizations adhere to the scale practices, others are seeking to learn from smaller organizations by employing scope strategies.

It is recognized widely that smaller organizations often possess the features of flexibility, responsiveness, innovativeness, product and process innovation, opportunity recognition, and niche satisfaction. Yet these are also features inherent in achieving economies of scope (Chandler 1990). There is mounting evidence that the more competitive larger organizations have begun seeing the light and altering their corporate strategy in line with economies of scope strategies. Some large biotech companies, particularly from the USA, but including companies such as Biocon and Mahyco in India, having not long ago been small themselves, see the value in working closely with the smaller biotech firms, and in some cases acquiring these companies. The implications for small biotech firms is that their exit strategies will in future be more likely to be the trade sale/merger option than the IPO option, however time will be the judge of the veracity of this prediction.

THE ORGANIZATION LIFE CYCLE AND THE PRODUCT LIFE CYCLE

The organization life cycle is therefore linked to the product life cycle given its focus on the way an organization grows in employee size and therefore management complexity. It is fundamentally different however in one important dimension – whereas the PLC is focused on the macro product market, the OLC is focused on the activities of a single productive unit. The life cycle of this single unit, as has been shown, is dependent on external influences.

The classic OLC starts with an innovative firm entering a new market where there is little or no competition. This is counterbalanced by a lack of organizational learning. New firms have no knowledge from past actions to draw on (Dodge et al. 1994, p. 125). As a market begins to emerge and expand through demand, so the need for an increase in productive capacity arises, bringing with it additional employee numbers and the need for new management competencies (Steinmetz 1969). Increasing profitability also signals opportunity for other firms. Existing businesses provide the working organizational models to follow. New entrants also bring alternate ideas with them. Through a process of variance and selection, better and better practices evolve (Aldrich 1979). However, as the market niche matures, resources begin to be depleted (markets start to dry up, inputs become scarcer) and competition increases. The shrinkage of profits leads to a focus on costs and economy of scale strategies (Dodge et al. 1994). Thus, the late stage of the OLC is characterized by intense competition in the environment and a shrinking market size.

Inevitably, if the OLC and PLC are linked, that is, an organization commits itself to a single product class, which follows the PLC, then the organization has a life span equivalent to the product. It grows and inevitably dies. Avoiding death (turnaround) is a process of moving to other PLCs (Churchill and Lewis 1983). Thus, the PLC describes the way a market niche develops, and the OLC the way an organization develops in its niche. This is often the problem of extreme niche marketing that is commonly seen in biotechnology in which, for example, an element of peptide analysis is targeted (in a business not scientific sense) by companies. The focus is make or break, if there are no breakthroughs or if the outcomes of the product development processes are not commercially viable, then the future of these companies will be limited. The discussion and diagram (Figure 7.5) which follow attempt to link the two in a common niche at an aggregate level in a life cycle of a population of organizations.

The desire is to develop increasingly profitable product life cycles which will create growth for the biotech company and achieve sustainability. Long term sustainability will be more difficult to achieve by many biotechnology companies that see IPO as their exit strategy, or as a worse case, see licensing of their initial product as their exit strategy.

In many industries technological developments have increased the diffusion of innovations through improved speed and quality of communication, improving market knowledge of both producers and consumers. The Internet, e-mail and access to generic technologies (such as commercial off-the-shelf technology), have all contributed to the speed at which ideas and products spread. However the main issue is that for biotechnology, due to high capital costs, technology diffusion will be slow without significant levels of investment. This investment tends to come either from

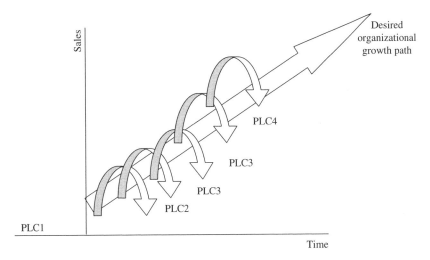

Figure 7.5 Optimal growth trajectory through multiple product life cycles

Big Pharma or from public and private investors. As opposed to the IT industry where $2–3 million would be required to get software designed, developed and on to the market within a six to twelve-month period; the extent of the biotechnology pipeline should be quite clear by now.

Furthermore, there is a very distinct difference between heroic/basic/blue sky research which is largely undertaken by public organizations such as universities, government agencies and research institutes, and incremental/ applied research and development which private sector firms are more likely to be involved in. Big Pharma are increasingly outsourcing their R&D (an increase from 5 per cent to 25 per cent in the last decade), despite the fact that they are also conducting more R&D, as they merge and seek to concentrate on market sales (Andrews 2002). Such a trend indicates that these companies recognize the long-term viability and importance of the research and diffusion of innovation process to new product development. A point which is missed by short-term, market-driven time horizons which are at odds with these industry structural characteristics.

Product diffusion is a problematic area for many biotech firms. Realistically most are research-oriented firms surviving on their inventiveness and innovativeness, based upon the quality of the science and originality of the research, usually indicated by the extent and value of intellectual property such as patents. The manufacturing environment is not a facet of business most are adept in. In fact many small biotech firms, the type seeking to list or recently listed, choose to stick to the R&D and license their IP to other manufacturers for production. This allows the biotech firm to focus on their core business, R&D, particularly as such licensing deals can be lucrative. By way of example, Genetic Technologies (GTG) a small biotech company has recently licensed its non-coding DNA patents to Myriad Genetics. An initial payment of $1.85 million was paid to GTG permitting it to expand its research agenda and the company has grown dramatically since (Trudinger 2002).

For the larger and even many medium biotech firms this is not really an option. Not only does the firm invent products, it must also bring them to market to ensure the cash flow that will fund the 'innovation stream'. The dilemma is firstly for firms to work out their strategic position and where their core business lies along the value chain. If they are more than an R&D firm then they must manage product development and product diffusion effectively. This therefore requires significant capital resourcing and understanding by investors of the long time horizons involved in the industry. If expectations are for rapid growth and return on investment, as in IT, then these expectations will not be met; funding will dry up, further slowing the diffusion of innovation process.

CONCLUSION

As more competitors move into the biotechnology industry, seeing the opportunities which have presented themselves and based upon, in most cases good science, the product life cycles for those candidates which do make it through an extended pipeline face the possibility of a short period of grace free of competitors. Biota's experience is testament to this.

With shortening product life cycles, shorter R&D cycles are essential. It is unclear whether there is much scope to continue to shorten these cycles.

These cycles cannot be viewed in isolation. In the next chapter the R&D, product, and organizational life cycles are merged with the broader cycles facing markets and the industry as a whole. To understand what is facing the NBF requires an analysis of all these cycles and a grasp of the impact of each.

REFERENCES

Aldrich, H. (1979), *Organizations and Environments*, Englewood Cliffs, NJ: Prentice-Hall.

Aldrich, H.E. (1990), 'Using an ecological perspective to study organisation founding rates', *Entrepreneurship Theory and Practice*, **14** (3), 7–24.

Andrews, P. (2002), 'The state of biotechnology in Australia', University of Queensland lecture, Brisbane, University of Queensland

Bird, B. (1989), *Entrepreneurial Behaviour*, London: Scott, Foresman.

Champion, D. (2001), 'Mastering the value chain', *Harvard Business Review*, June, 109–15.

Chandler, A. (1990), *Scale and Scope: The Dynamics of Industrial Capitalism*, Cambridge, MA: Harvard University Press.

Chaston, I. (1997), 'Small firm performance: assessing the interaction between entrepreneurial style and organizational structure', *European Journal of Marketing*, **31** (11/12), 814–25.

Chell, E. (1996), 'Culture, entrepreneurship and networks revisited', *ICSB 41st World Conference Proceedings*, 16–19 June, Stockholm.

Churchill, N. and V. Lewis (1983), 'The five stages of small business growth', *Harvard Business Review*, **61** (3), 30–50.

Coase, R. (1937), 'The nature of the firm', *Economica*, **4**, November, 386–405.

Dodge, H. and J. Robbins (1992), 'An empirical investigation of the organisational life cycle model for small business development and survival', *Journal of Small Business Management*, **30** (1), 27–37.

Dodge, H., S. Fullerton and J. Robbins (1994), 'Stage of the organisational life cycle and competition as mediators of problem perception for small business', *Strategic Management Journal*, **15** (2), 121–34.

Emery, F. and E. Trist (1965), 'The causal texture of organisational environments', *Human Relations,* **18**, 21–32.

FDA (2004), Available: http://www.fda.gov/bbs/topics/NEWS/2004/NEW01005.html, Accessed 12 July 2004.

Greening, D., B. Barringer and G. Macy (1996), 'A qualitative study of managerial challenges facing small business geographic expansion', *Journal of Business Venturing*, **11** (4), 233–56.

Greiner, L.E. (1972), 'Evolution and revolution as organizations grow', *Harvard Business Review*, **50** (4), 37–46.

Helms, M. and T. Renfrow (1994), 'Expansionary processes of the small business: a life cycle profile', *Management Decision*, **32** (9), 43–6.

Hine, D. and A. Griffiths (2004), 'The impact of market forces on the sustainability of the biotechnology industry', *International Journal of Entrepreneurship and Innovation Management*, **4** (2/3), 138–54.

Howard, D. and D. Hine (1997), 'The population of organisations life cycle', *International Journal of Small Bussiness*, **15** (3), April–June, 30–41.

Jeffress, P. (1991), 'Achieving change through organisational entrepreneurs', *Asia-Pacific International Management Forum*, **17** (1), 13–20.

Kessler, E. and P. Bierly (2002), 'Is faster really better? An empirical test of the implications of innovation speed', *IEEE Transactions on Engineering Management*, **49** (1), 2–22.

Kessler, E. and A. Chakrabarti (1996), 'Innovation speed: a conceptual model of context, antecedents, and outcomes', *Academy of Management Review*, **21** (4), 1143–92.

Klofsten, M. (1994), 'Technology-based firms: critical aspects of their early development', *Journal of Enterprising Culture*, **21**, 535–57.

Love, J. and B. Ashcroft (1999), 'Market versus corporate structure in plant-level innovation performance', *Small Business Economics*, **13** (2), 97–107.

Miller, D. and P. Friesen (1984), 'A longitudinal study of the corporate life cycle', *Management Science*, **30** (10), 1161–83.

Rogers, E. (1995), *Diffusion Of Innovations*, 4th edn, New York: Free Press.

Schumpeter, J. (1939), *Business Cycles: A Theoretical, Historical and Statistical Analysis of the Capitalist Process*, New York: McGraw-Hill.

Slevin, J. and D. Covin (1990), 'New venture strategic posture, structure, and performance: an industry life cycle analysis', *Journal of Business Venturing*, **5** (2), 123–35.

Specht, P. (1993), 'Munificence and carrying of the environment and organisation formation', *Entrepreneurship Theory and Practice*, **17** (2), 77–86.

Steinmetz, L.L. (1969), 'Critical stages of small business growth: when they occur and how to survive them', *Business Horizons*, February, 29.

Trudinger, M. (2002), 'Jacobsen defends deal', *Australian Biotechnology News*, **1** (38), 1–5.

Watts, G., J. Cope and M. Hulme (1998), 'Ansoff's Matrix, pains and gain. Growth strategies and adaptive learning among small food products', *International Journal of Entrepreneurial and Research*, **4** (2), 101–16.

Williamson, O.E. (1985), *The Economics of Capitalism: Firms, Markets, Relational Contracting*, New York: Free Press/Collier Macmillan.

8. The cycle game II – business, market and industry cycles

INTRODUCTION

This chapter will discuss industry and business cycles from the context of the firm unit, including innovation diffusion and product diffusion models, market life cycles and industry life cycles, opportunity recognition and growth through cycles. Life cycle models of products, firms and markets have proved to be a rich source of microeconomic research activity since the 1960s, when they were posited by researchers such as McGuire (1963) and Vernon (1966). We explore some alternative life cycle analyses in this chapter and seek to understand the extent to which they apply in the biotechnology industry as well as understanding the uniqueness of biotechnology in its own cycles and industry development.

While evidence is clear that R&D spend in both pharmaceuticals and in biotechnology is up, even for Big Pharma, there is also sufficient evidence that increasingly this work is being completed by smaller biotechnology companies. Not only are Big Pharma outsourcing their R&D in many cases to biotechnology companies, the result of mergers and conglomeration is their increased focus on market sales and distribution channels. This has created a fresh opportunity for biotechnology companies with R&D spend increasing, despite drug approvals faltering. Such a trend indicates that these smaller biotechnology companies recognize the long-term viability and importance of research and development and new product development. This point is missed by short-term, market-driven time horizons that are at odds with the structural characteristics of the biotechnology industry which are geared very much for the long term.

The chapter argues that if the tenet established in the previous chapter, that business is impacted by cycles, holds, then it is not sufficient to concentrate on one or two types of life cycles. If all life cycles are linked, the biotechnology entrepreneur must consider not only the impact of each life cycle individually, but also the interactions between them. With life cycle analyses, we extend from the micro focus on the product and firm to broaden out to the entire biotechnology industry.

INNOVATION AND INDUSTRY

Schumpeter in his writings of the industrial era wrote extensively about the creative destruction of innovation, particularly at the industry level. Innovation itself can be observed at various levels. The work of Schumpeter, Arrow and Williamson, largely commented on the industry level impacts of innovation. Others such as Burns and Stalker (1961), Normann (1971), Lawrence and Lorsch (1967) and Oakey (1984) have concentrated on the firm level impacts and antecedents of innovation. Others still (Baum and Haveman 1997) have chosen to observe both firm level and industry level impacts of innovation and change.

Abernathy argues that product innovation predominates early in the life of an industry where the product design is still in flux, so that economies of scale are not attained and the production system is inefficient. However as the dominant design is established and standardized process innovation takes over, so that cost and productivity become much more important in a mature industry with well developed products, so price competition becomes the dominant paradigm (Abernathy and Utterback 1978).

Tushman and Anderson's (1986) work, which uses data from the mini-computer, cement and airline industries from inception until 1980, demonstrates that 'technology evolves through periods of incremental change punctuated by technological breakthroughs that either enhance or destroy the competence of firms in an industry'. Importantly these researchers' findings indicated that the firms that initiate major technological change and progress grow more rapidly than other firms.

Such writings have given rise to an extension of the innovation and change literature in recent years and the exploration of partial equilibrium and continuous change (Brown and Eisenhardt 1997), bodies of literature, which assume intermittent upheavals in the technological trajectories of markets and industries. These approaches are closely aligned, with their lineage being easily traced to Schumpeter's creative destruction concepts.

CHANGE AS A CONSTANT

High technology industries face extraordinary rates of change in terms of technical change, as well as changing business conditions. The external environmental influences, which create change in the biotechnology industry, are more complex than in most industries given that much of the biotechnology industry is about human intervention. The social, ethical and religious consideration of biotechnologies including stem cell research, genetic engineering such as cloning, genetically modified organisms and genetically modified foods have not only created debates in social and religious sectors of the community but have major economic impacts such as stem cell moratoria in the USA which have actually created opportunities for companies in countries such as Singapore where ES Cell, and embryonic stem cell supply company saw a rapid increase in the demand for their product. So too did another Singaporean company CordLife Pty Ltd. CordLife was initially setup as the first private cord-blood bank in Singapore with a facility to freeze stem cells at birth. However the vision for CordLife has expanded to become the leading stem cell biotechnology company offering highest quality cellular banking, R&D and regenerative medicine in Asia. In effect a religious and moral decision made in the USA has spawned an industry in Singapore. To further promote stem-cell research, Singapore held the inaugural International Stem Cell Conference (ISCC) in October 2003. As one of the largest stem cell meetings to date, this event was attended by more than 500 clinicians and scientists from around the world, and discussed and debated the latest in basic and applied stem cell technologies.

Developments such as this are perennial in the industry. The external environment is turbulent and complex. It is almost inevitable then that change is something biotechnology companies face constantly. This differs from the traditional concept of change in the industrial setting as developed by Schumpeter and pursued in Schumpeterian evolutionary economics. Figure 8.1 below indicates the type of cyclic change patterns envisaged by Schumpeter.

Figure 8.1 represents the situation which has traditionally occurred in the industrialized sector. In the work of Schumpeter (1934), the institutional framework had a major role in the industry innovation processes. The impact of developments in innovative capacity were infrequent and devastating to the existing industry. The industry would move through a period of turmoil during the innovative upheaval in which existing firms would be displaced or dramatically changed. This is the process of creative destruction referred to by Schumpeter. However over time this innovative upheaval would be integrated to the dramatically changed industry. As this occurred, the rate of change in the industry would reduce to a near zero level until the next innovative upheaval. The extent of the upheaval would be impacted by institutional factors which would promote or reduce the diffusion of the innovation.

The shift in focus of the biotechnology industry can be taken as a constant. This takes the industry's development into a neo-Schumpeterian realm, where change is pervasive.

Figure 8.2 above portrays a different situation, one which is more likely to face new and emerging industries, particularly those in sectors of the

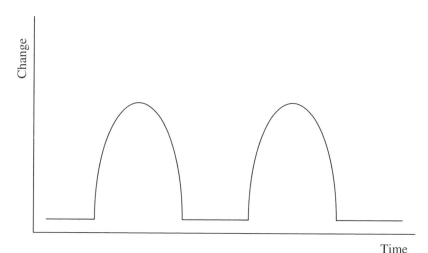

Figure 8.1   Traditional Schumpeterian change in traditional industries

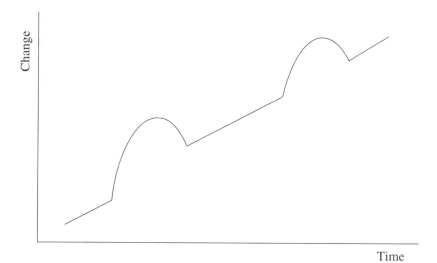

Figure 8.2   Dynamic change in new economy industries

economy more directly impacted by changing and developing technology. Due to the constant development and implementation of new science and new technology, as promulgated through scholarly publications and patents, most areas of biotechnology operate in a climate of near constant change. However even in this change climate there come opportunities for groundbreaking innovations which will create more dramatic change fluctuations. Examples of this in biotechnology include most obviously, recombinant DNA, and then there are PCR, the completion of the mapping of the human genome, cloning of Dolly the sheep, to name but a few.

In biotechnology the institutional framework serves to support the diffusion of innovation throughout the industry, through knowledge sharing, licensing deals, publications and the purchase of new technology particularly through public funding. The height of the fluctuations is a function of the extent of support or hindrance of the institutional framework surrounding the innovation. A supportive institutional environment will create higher peaks in what could still be regarded as the creative destruction phase of industry development. However the destruction will not be nearly as significant as firms within the industry are more used to and willing to incorporate change and the innovations which emerge. For example, polymerase chain reaction (PCR), invented by Kary Mullins in 1982, took a number of years of experimentation and development to be made commercially available. However because it achieves rapid DNA replication, it has been diffused widely amongst the scientific community. The impact of PCR is similar to many innovations, it streamlines the production process. It also makes the genetic testing process much quicker and easier. In fact as a development it was so important that Roche paid Cetus Corporation, where Mullins made the discovery, $300 million for the IP and Mullins received the 1993 Nobel Prize for Chemistry for the discovery. So in this case the opportunity for creative destruction was ameliorated by the rapidity and ease of diffusion of the innovation in a receptive industry sector.

In Figure 8.2, the slope of the line represents the extent of incremental (evolutionary) innovation occurring between the revolutionary peaks. The institutional framework can also have an impact upon the slope of the change line, in its degree of support for or hindrance of innovative efforts within the industry. However the impact of the institutional framework will be mitigated by other industry specific factors such as propensity to adopt new technology, ability to incorporate generic technological developments into the industry and diffusion of industry developed innovations.

As a result of the barrage of new technologies and scientific developments in the biopharma area, the typical biotechnology company has morphed as well. The shift has been from product companies with DNA and molecular based techniques (Amgen stands for Applied Molecular Genetics Inc.), to tool-based technologies enabled by chip transistorization and then platform technologies with multiple applications.

Aldrich (1990) made the observation that a specialty, technology-based, innovation is often linked to entrepreneurship and business foundings.

Most innovations are confidence enhancing rather than confidence destroying (Schumpeter 1939; Kirchhoff 1996; Schmookler 1966; Tushman and Anderson 1986) and can thus be adopted by existing organizations. Confidence-enhancing innovations are order of magnitude improvements in price/performance that build incrementally on existing know-how within a product to service class. These are incremental innovations that incumbents can incorporate into their own processes relatively easily. In contrast confidence-destroying innovations require new skills, abilities and knowledge in the development or production of a product or a service. These are the truly radical innovations Abernathy and Utterback (1978) saw as leading to discontinuities. These types of innovations fundamentally alter the competencies required of an organization as they revolutionize the product market.

Therefore all innovations have an impact on the competitiveness of individual firms. Adoption rates of new technology such as PCR are an important lead indicator of future competitiveness. The new technology is expensive, for example a 900MHz Nuclear Magnetic Resonance (NMR) machine, used for NMR spectroscopy at the atomic level will cost around $10 million. Not many universities, medical research institutes or biotechnology firms let alone NBFs can afford this capital outlay. Therefore in the adoption of new technology in biotechnology, size and institutional support are important. Economies of scale still come into play in a capital-intensive industry such as biotechnology, and early adoption means a quicker infusion of the benefits afforded by technological advances.

The scale imperative is explained by Chandler in recalling that 'in order to benefit from the cost advantages of these new high-volume technologies of production', entrepreneurs had to make three interrelated sets of investments. The first was an investment in production facilities large enough to exploit a technology's potential economies of scale or scope. It was this three pronged investment in production, distribution and management that brought the modern industrial enterprise into being (1990, p. 4). Scale was not only directed at production but at distribution and in a similar way management, in which replication of standardized practices created the volume advantage. This helps also explain to some extent the conglomeration of major biopharmaceutical players.

However, the confidence-enhancing and confidence-destroying developments in the biopharma industries has had a major impact on the number and size of players. While this will be explored in more depth in Chapter 12, it is worthwhile undertaking a comparison of the number of major players over time. Tables 8.1 and 8.2 list members of the Pharmaceutical Manufacturers Association (PMA) at two points in time: 1988 and 2003.

Table 8.1   1988 PMA Members (42 players)

• Abbott Laboratories	• G.D. Searle	• Procter & Gamble
• American Cyanamid	• Glaxo SmithKline	• Rhone Poulenc
• A.H. Robins	• Hoechst	• Rorer
• Asatra	• Hoffman-LaRoche	• R.P. Scherer
• BASF	• ICI	• Roussel
• Beecham Laboratories	• Johnson & Johnson	• Sandoz
• Boehringer Ingelheim	• Knoll	• Schering-Plough
• Boots Pharmaceuticals	• Eli Lilly	• SmithKline Beecham
• Bristol-Myers	• Marion Laboratories	• Squibb
• Carter-Wallace	• Merck	• Sterling Drug
• Ciba Geigy	• Merrell Dow	• Upjohn Company
• Connaught Laboratories	• Monsanto	• Warner-Lambert
• DuPont Pharmaceuticals	• Pfizer	• Wellcome
• Fisons Corporations	• Pharmacia	• Zeneca

Source: Greg Orders, Progen Industries

Table 8.2   2003 PMA Members (16 players)

• Abbott Laboratories	•	• Proctor & Gamble
• American Home Products	• Glaxo SmithKline	•
•	•	•
• AstraZeneca	• Hoffman-LaRoche	•
• Aventis	•	•
• BASF	• Johnson & Johnson	•
• Boehringer Ingelheim	•	• Schering-Plough
•	• Eli Lilly	•
• Bristol-Myers Squibb	•	•
•	• Merck	•
•	•	•
•	• Novartis	•
•	• Pfizer	•
•	•	•

Source: Greg Orders, Progen Industries

The tables show a dramatic decline in the number of major players in the market as they have merged and been taken over, or have left the association.

This is indicative of a consolidation trend in biopharmaceuticals. However, it does not necessarily have immediate implications for other smaller players in the industry, unless of course the strategies of these con-glomerated companies shift towards a broader product and R&D base. Evidence of this in pharmaceuticals is not apparent as yet.

SCALE AND SCOPE

As we have seen in Chapter 7 there are competing influences on the future directions of the biotechnology industry and the likely dominance of small or large firms. Confidence-enhancing innovations do not wipe out whole generations of large firms, however, as can be seen in the PMA membership in Tables 8.1 and 8.2.

POPULATION ECOLOGY LENS

Another way to view the life cycles of markets and industries is through the population ecology lens. One of the current research themes of the life cycle approach is focused on understanding industry life cycles and the way busi-ness populations evolve within different industrial settings.

Aldrich (1979) proposed a generic model of how such populations develop, based on principles from ecology. Aldrich proposed stages of pop-ulation density dependent upon the level of competition for resources in a niche. A niche, in this context, is a strategic business location or resource space, while resources are defined generically in terms of scope for product market growth, access to inputs and availability of relevant technology.

Niches change over time as product markets grow/decline or technologies change, so they vary in attractiveness. A rich niche encourages new businesses to form and existing businesses to grow, and the tendency is for population density to increase in rich niches. In contrast, in the declining niche, the busi-ness population falls. A niche's carrying capacity in terms of the number of businesses is determined then, by a combination of two main forces, the increasing legitimacy of business formation/growth that attaches to a new niche because of resource abundance, followed by diminishing attractiveness of the niche as competition increases for depleted resources (Specht 1993).

Since 1995, population ecology research has been criticized for its almost exclusive focus on the forces in the niche (the external environment) and its neglect of the decision-making processes that take place inside firms (Baum and Haveman 1997). It is possible to address this criticism, in part at least, by using empirical data to link strategy within the firm to stages of niche development.

Howard and Hine (1997) proposed a generic four-stage model (based on population ecology principles), to describe the process by which the environment changes as an industry develops, namely, the population of organizations life cycle (POLC). The four stages were defined as:

New niche formation

A new niche forms around innovation in product design and/or new technologies. One or a very few firms set up and attempt to define and exploit this innovation. A new industry starts, the product market starts to grow, suppliers begin organizing and the production techniques start to develop.

Emerging niche

The niche starts to mature as the market grows. The success of the pioneers is noticed and imitated, increasing the rate of new business formation and the niche becomes a popular place for enterprises to be; it becomes legitimized. Thus, population density in the niche grows as does competition for resources, resulting in a slowing of the rate of new organization formation.

Mature niche

In the mature niche, market growth has slowed as a steady state is approached. The size structure of the business population has been rationalized in terms of a fit that fully exploits the available resources. New business formation normally displaces others. Standardization in marketing, product design and production techniques encourage economies of scale and the tendency for average business size to increase.

Declining niche

A niche in decline reflects saturated product markets. The number of organizations in the niche declines as obsolete technologies wear out. Competition for what is left of product markets may drive organizations from the niche to more attractive ones. Average business size in the niche may increase.

Table 8.3 summarizes the POLC model in terms of the major characteristics of the niche environment, given its stage of development, in terms of marketing and production functions.

The four stages describe an industrial life cycle in terms of the changing external environment and hence the context for internal strategies. For example, in the early stages of new niche development, market potential may be huge given limited competition, but the market will be unaware of

Table 8.3   POLC stage environmental characteristics

Enterprise function	New niche	Emerging niche	Mature niche	Declining niche
The firm's markets	Potential undefined	Product variants define the market	Product standardized and market saturated	In decline
Production technology	Scope for different patterns of organizing and producing	Organizational forms stabilizing and better practices emerging	Best practices defined and economies of scale developed	Organizational variation potential virtually nil
The firm's supply	Innovative sourcing required	Supply becoming stable and standardized	Supply standardized	Suppliers moving to new niches

the firm's new product. Logically, a particular strategy or modus operandi is the appropriate response for effectively exploiting the state of the niche. A new niche would require innovative management approaches to niche exploitation (Howard and Hine 1997). In Thompson's (1999) schema, a charismatic role for the entrepreneur would be recommended because it leads to creative solutions.

Later on in niche development, as the niche matures, resources such as product markets and production inputs become thoroughly defined and standardized, suggesting the need for efficiency in resource utilization. In formal planning terms, the enterprise might implement a best practices plan. Again, if the internal strategy is to 'fit', then Thompson's architectural role would seem appropriate.

COMPETING SCIENTIFIC AND MARKET FORCES IN THE INDUSTRY

The forces impinging on the growth and development of the industry can be loosely located into two camps: first, the scientific and technical forces often associated with technology push innovation from scientists and researchers and technical specialists in biotechnology. Evidence of performance and development in this area is indicated by the intellectual property which is an outcome of their efforts. In biotechnology this is most clearly evidenced through two means: patents and publications. In this realm, diffusion of knowledge does necessitate a corresponding desire to commercialize. Second, the market forces and the desire to commercialize ideas and research through product innovation, providing a revenue stream and a return to investors also has a major influence on the development of the industry.

In many industries, technological developments have increased the diffusion of innovations through improved speed and quality of communication, improving market knowledge of both producers and consumers. The Internet, e-mail, access to generic technologies (such as commercial, off-the-shelf technology), have all contributed to the speed at which ideas and products spread. However the main issue is that for biotechnology, due to high capital costs, technology diffusion will be slow without significant levels of investment. This investment tends to come either from Big Pharma or from public and private investors. These investors have considerable influence in the industry, and have the potential to alter the direction of the industry through their selective funding of scientific projects. For instance in the final chapter we look at globalization. In exploring the pattern of exports, subsidiary establishment and major collaborations of biotechnology companies, we come to question whether the phenomenon should be referred to as globalization or simply Westernization.

The patterns display a focus on the ailments which afflict wealthy western nations such as the USA, and Europe, with major funding directed at R&D in these areas, and a neglect of poorer regions such as South America and Africa. There is some evidence that biotechnology is about high-cost, high-value products being developed only for those who can afford them. This tends to differ from the public good objectives of most scientists and research institutions. Add to this the expectation of short-term gain by the typical investor and we see clear potential for market-driven tangents in the life of the biotechnology industry.

The science needs the injection of funds. However, investors, while still new to the biotechnology industry will expect short-term gains. Where the gains are not made many investors may shift their view toward other industry sectors where their 30 per cent return can be guaranteed. It would be dangerous to withdraw investment funding in these early stages of development for most biotechnology industries around the world, where, more than ever before, it can be argued that the industry is on the brink of making a commercial return for investors. However, the short-term demand on returns made by market-driven polices and practices prove to be disruptive rather than enhancing the long-term sustainability of the industry. While market capitalization is important to this industry, so too is its scientific base. Market-driven cycles can only serve to exacerbate a split between biotechnology and its scientific base – where capability activities should be seeking to build this (Hine and Griffiths 2003).

Further, financial benchmarks are critically important as a basis for judging investment decisions for listed companies. Yet how can we have a price to earnings ratio for a company, which has no earnings? Put simply many of the biotech ventures don't have sales revenues. Down the track when their licences start producing progress payments there will be something to consider. Also where the venture undertakes such commercial tasks as contract research there will be revenue to consider.

The sustainability of an industry should be indicated by the level of innovation occurring in the industry, demonstrating the likely product stream that will maintain the growth path of the industry. In biotechnology this product stream is indicated by both scientific publications and its commercial parallel, intellectual property, most commonly indicated by patents. Tracking patents, as well as the state of the major biotechnology business indicator, the NASDAQ Biotechnology Index, can provide an idea of which of the forces is taking the lead, science or industry. This work was conducted for a recent paper (Hine and Griffiths 2003) and is offered here to support our analysis.

Figure 8.3 indicates a more linear growth pattern in the science and scientific discovery than displayed in the fluctuating NASDAQ. This could be interpreted as an indication of the robustness of the science (as indicated by patent approvals by the US Patent and Trademark Office). Another perspective could be that the market fluctuations do not have a direct impact on the science, however, there may be an indirect impact. Patent counts are about future value and product streams that are yet to be commercialized

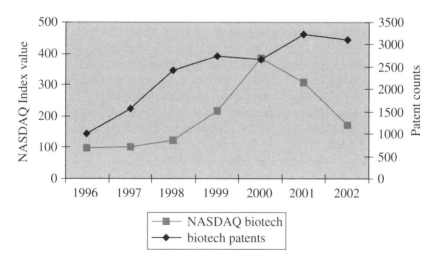

Figure 8.3   NASDAQ biotechnology index and biotechnology patent

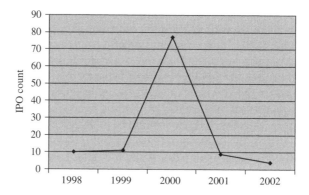

Figure 8.4 IPO biotechnology sector count

and therefore represent future value. While there were small hiccups in the progress of each patent growth path, the protection of intellectual property remains important despite market turbulence.

Another lead indicator of the state of play in the biotechnology industry is the number of floats (initial public offerings – IPOs) which occur each year, in this case on the NASDAQ. It is a measure of confidence in the industry. On the flip side, the number of IPO withdrawals, that is companies who have filed for listing, but have withdrawn.

Figure 8.4 indicates the enormous impact the tech crash had on IPOs. Influencing factors are many and varied, but put simply, investors just pulled their money out of the sectors, leaving no support in terms of subscription to share issues for the new floats to go ahead.

IPOs can be seen as a lead market indicator. As the tech crash occurred, the plummeting prevalence of IPOs provided dramatic evidence of industry confidence in the immediate future, as opposed to patents which indicate the consideration of a longer-term future, the start of an R&D process which may eventually culminate in a new product. After the crash, confidence took time to rebuild at the industry level, while patents proceeded unabated.

Figure 8.5 can be seen to be a proxy measure of the loss or gain in confidence in the industry using lead indicators.

The data in Figure 8.5 provides an indication of the volatility of the market forces on an industry which seeks to build its capacity and success rate over a long period of time. It is an industry, as we observed in Chapter 7, where it is very difficult to shorten R&D cycles and speed up the innovation process. It is difficult to hurry the science which underpins the industry. The fluctuations which have occurred at the market level have had an impact on the commercialization of the science. Funding sources wax and wane,

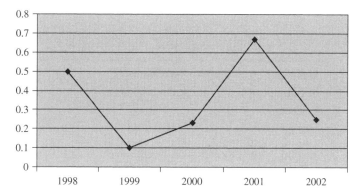

Figure 8.5   Proportion of IPO withdrawals from total IPOs listed

investors need to be educated about the industry. Where information asymmetry remains, the future for the industry will remain hazardous.

The influence that the two competing forces of market and science have on the level and type of development that occurs in the industry is clear from the previous analysis. To make sense of these competing forces and to permit some gauging of the health of the industry, we offer the resultant industry development matrix in Figure 8.6. This matrix combines market and scientific forces and indicates the differential influence they have on industry development (with evidential indicators supplied). While the competing forces are not mutually exclusive in their influence, a harmony needs to be attained so that both product and knowledge development occur simultaneously. Under such a harmonious condition both knowledge diffusion and product diffusion can occur. The potential impact of this is to enhance the sustainability of the industry. The harmonious condition is evident in explosive development. This creates a balance between market diffusion and science diffusion, which satisfies commercial and scientific imperatives.

In terms of the development of the industry, not all forms of development are positive. For sustainable development to occur requires the harmonization between the two forces. A dichotomy is evident between the two sides of the matrix, simplified in Figure 8.7. While the 'degenerative development' is indisputably unsustainable, involving a situation where no new ideas are being published, no new ideas are being patented, no new firms are emerging and no new products are emanating from the industry, the future would seem bleak. While not as dire, 'unhinged development' is certainly not going to be a sustainable industry development situation either. It is unsustainable to have high IPO rates, while there are few patents

**High
market
imperative**

Unhinged development Evidence: Low publications Low patents/IP High IPOs Low new products	Dynamic development Evidence: High publications High patents/IP High IPOs High new products
Degenerative development Evidence: Low publications Low to no patents/IP Low to no IPOs Low to no new products	Knowledge development Evidence: High publications High patents/IP Low IPOs/high withdrawals Low new products

Low   (left of table) **science imperative**

High   (right of table) **science imperative**

**Low
market
imperative**

Source: Hine and Griffiths 2003

Figure 8.6   The high technology industry development matrix

to establish and support an innovation stream upon which the product stream is based. The publication base has to be built up over time.

While market analysts, investors, venture capitalists and other market-focused stakeholders would be unhappy with the situation apparent in 'knowledge development', it is a situation that will sustain the industry into the future. The science and the patents from which new products will emerge are still growing in number and strength. This is reminiscent of what the biotech sector have been facing in recent years. While abandoned by investors for some time, patents have been maintained. Such a phenomenon suggests that the long-term outlook for biotechnology is bright, supporting its sustainability and reaffirming the critical importance of science as a force underpinning the industry.

Unsustainable development	Sustainable development
Unhinged development	Dynamic development
Degenerative development	Knowledge development

Source: Hine and Griffiths 2003

Figure 8.7   The development dichotomy

The matrix displays four quadrants. However this does not denote a cyclic progression. As high-technology industries develop, it is not essential that they move through each of the quadrants sequentially. For healthy sustainable high-technology industries however, it can be expected that there will be some modulation between the 'dynamic' and the 'knowledge development' quadrants, indicating the varying influences of market and science forces at different times of their development. It would be expected that only mature and declining industries will face the left-hand quadrants.

FUTURE SCENARIO

What appears to have been happening in biotechnology in recent years has indicated the competing and possibly complementary nature of the two forces on industry development. While we do not dare to extend this argument to a Nash equilibrium analysis, it would clearly be desirable that these competing forces are harmonized to create an equilibrium state to maximize developmental potential, as in the 'dynamic development' quadrant or even the 'knowledge development' quadrant. However, the evidence of recent years has pointed more to the disequilibrium state, in which dramatic fluctuations have occurred particularly in market forces that have impinged on the development of biotechnology. It seems likely that these dramatic fluctuations mitigate over time as investors, financiers and industry analysts all improve their own knowledge base of the industry and develop more mature, focused, perspectives, rather than generic broad-brush investment criteria. This enhanced knowledge base amongst the market-oriented stakeholders will be indicated when dedicated investment financial ratios are developed for biotechnology reflecting its R&D cycles, funding requirements, capital base and product life cycle. The sound basis for judgement on the status of each industry and their associated companies will lead to fluctuations in line with other market measures such as the

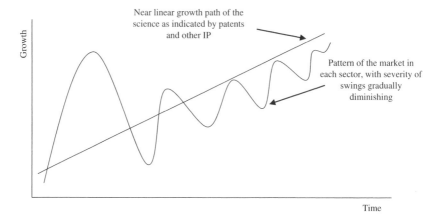

Figure 8.8   Future scenario

Dow Jones, the DAX and the FTSE100. This harmonizing process which would occur over time, may be reflected in the indicative 'future scenario' in Figure 8.8.

There are several important lessons to be drawn from this analysis for biotechnology. The first to note is that biotechnology as an industry is maturing, oscillating between 'dynamic' and 'knowledge' quadrants, but showing little evidence of a slide into 'unhinged' or 'degenerative development'. There is evidence of a shift along the value chain for the biotechnology industry, from an emphasis on discovery research to development, manufacture and marketing. This was identified by Mark Levin CEO of Millennium Pharmaceuticals (Champion 2001, p. 109) in his observation that 'value has started to migrate downstream, towards the more mechanical tasks of identifying, testing, and manufacturing molecules that will affect the proteins produced by genes, and which become the pills and serums we sell'. This shift in the creation of value along the value chain will continue to occur as more companies move through their trial phases and approval processes towards the market. This commercialization push is also supported by the number of drug approvals indicated by the FDA's Center for Drug Evaluation and Research (CDER 2002).

Much of the empirical evidence shows us that generating sustainable competitive advantages for firms, industries and national economies is not simply achieved by the creation and substitution of new technologies (such as the Internet) for old. Sustainable strategies require the development of industry (Porter 2001) and firm capabilities (Mathews 2000). Evidence suggests that management fads make it difficult to discern when and how companies engage with new practices (Micklethwait and Wooldridge 1997).

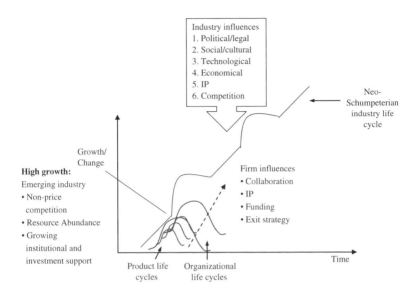

Figure 8.9   The biotechnology industry life cycle

CONCLUSION – THE BIOTECHNOLOGY INDUSTRY LIFE CYCLE

In conclusion, Figure 8.9 provides a summary of the interaction of the critical life cycles in the development of the biotechnology industry.

Each cycle impinges on the other and can't be considered in isolation. Change is dynamic and constant, as the external influences are potent. Responding to these external influences requires constant innovation within the biotechnology firm, placing pressure on the R&D cycle and the pipeline. An innovation stream leads to a product stream, through this the NBF can build an IP profile which will permit it to undertake deals which will assist its organizational growth. However, none of this is possible if the industry is not being supported so as to achieve a degree of sustainability. If the industry is starved of resources because of faddish investment decisions then all firms in the industry suffer.

REFERENCES

Abernathy, W.J. and J.M. Utterback (1978), 'Patterns of industrial innovation', *Technology Review*, **80**, June–July, 2–9.
Aldrich, H.E. (1979), *Organisations and Environments*, New Jersey: Prentice Hall.

Aldrich, H.E. (1990), 'Using an ecological perspective to study organisation founding rates', *Entrepreneurship Theory and Practice*, **14** (3), 7–24.

Baum, J. and H. Haveman (1997), 'Love thy neighbor? Differentiation and agglomeration in the Manhattan hotel industry, 1898–1990', *Administrative Science Quarterly*, **42** (2), 304–38.

Brown, S. and K. Eisenhardt (1997), 'The art of continuous change: linking complexity theory with time-paces evolution in relentlessly shifting organizations', *Administrative Science Quarterly*, **42**, 1–34.

Burns, T. and G. Stalker (1961), *The Management of Innovation*, London: Tavistock.

Champion, D. (2001) 'Mastering the value chain: an interview with Mark Levin of Millennium Pharmaceuticals', *Harvard Business Review*, **79** (6), 108–15.

Chandler, A. (1990), *Scale and Scope: The Dynamics of Industrial Capitalism*, Cambridge, MA: Harvard University Press.

Coase, R. (1937), 'The nature of the firm', Economica, **4** (Nov.), 386–405.

Eisenhardt, K. and S. Brown (1998), 'Timepacing: Competing in markets that won't stand still', *Harvard Business Review*, **76** (2), 59–69.

Hine, D. and A. Griffiths (2003), 'Sustainability of the new economy: cautionary lessons of 'e' fads', *International Journal of Management and Decision Making*, **4** (/23), 230–46.

Howard, D. and D. Hine (1997), 'The population of organizations life cycle (POLC): implications for small business assistance programs', *International Small Business Journal*, **15** (3), 30–41.

Kirchhoff, B. (1996), 'Self-employment and dynamic capitalism', *Journal of Labor Research*, **17** (4), 627–43.

Lawrence, P. and J. Lorsch (1967), 'Differentiation and integration in complex organizations', *Administrative Science Quarterly*, **12**, 1–30.

Mathews, J. and C. Dong-Sung (2000), *Tiger Technology: The Creation of a Semi-Conductor Industry in East Asia*, Cambridge: Cambridge University Press.

McGuire, J.W. (1963), 'Factors affecting the growth of manufacturing firms', Bureau of Business Research, University of Washington.

Micklethwait, J. and A. Wooldridge (1997), *The Witch Doctors: What the Management Gurus are Saying, Why it Matters and How to Make Sense of It*, London: Mandarin.

Normann, R. (1971), 'Organisational innovativeness: product variation and reorientation', *Administrative Science Quarterly*, **16**, 203–15.

Oakey, R. (1984), *High Technology Small Firms: Regional Development in Britain and the United States*, London: Frances Pinter.

Porter, M. (2001), 'Strategy and the Internet', *Harvard Business Review*, **79** (3), 62–78.

Schmookler, J. (1966), *Invention and Economic Growth*, Cambridge, MA: Harvard University Press.

Schumpeter, J.A. (1934), *The Theory of Economic Development*, Cambridge, MA: Harvard University Press.

Schumpeter, J. (1939), *Business Cycles: A Theoretical, Historical, and Statistical Analysis of the Capitalist Process*, New York: McGraw-Hill.

Specht, P. (1993), 'Munificence and carrying of the environment and organisation formation', *Entrepreneurship Theory and Practice*, **17** (2), 77–86.

Thompson, J. (1999), 'A strategic perspective of entrepreneurship', *International Journal of Entrepreneurial Behaviour and Research*, **5** (6), 279–92.

Tushman, M.L. and P. Anderson (1986), 'Technological discontinuities and organisational environments', *Administrative Science Quarterly*, **31** (3), 439–65.
Vernon (1966), 'International investment and international trade in the product cycle', *Quarterly Journal of Economics*, **80**, May, 190–207.
Williamson, O.E. (1985), *The Economics of Capitalism: Firms, Markets, Relational Contracting*, New York: Free Press, Collier Macmillan.

9. Public policy, regulatory and ethical challenges facing the entrepreneurial biotechnology firm

INTRODUCTION

It is often assumed that the influence of the external environment is only one-way, with external environmental factors considered as independent variables. This is rarely the case in the biotechnology industry. Even for small biotechnology businesses there is scope to influence the environment the firm is operating in, though not usually in isolation.

This chapter categorizes and evaluates the major public and public policy influences on the entrepreneurial biotechnology firm and their ability to innovate, such as: the extensive regulatory environment in biotechnology, as exemplified by the complex and long drug approval process for FDA and EMEA; the nature of a global industry and the impact of the changing corporate strategies of the big pharmaceuticals. Fundamental to this analysis is consideration of ethical frameworks and motivations for establishing codes of ethics and meeting ethical standards. In many instances, ethics can be used as a marketing tool, as can regulatory compliance.

This chapter takes a broad view of the philosophical underpinnings of ethics and ethical decisions as applied to bioethics. The role ethics plays as an enhancer or hinderer of the innovation process for young biotechnology firms is highlighted through case examples. External ethical perspectives on the biotechnology industry are considered, as portrayed in national studies such as that by Biotechnology Australia (2005) on public perspectives of biotechnology.

THE REGULATORY ENVIRONMENT

Public policy significantly affects the entrepreneurial biotechnology firm and its ability to innovate.

Few industries are so affected by the four external environmental factors as biotechnology. These are socio-cultural factors with regard to ethics, which link in strongly with political/legal factors such as regulation. These are mixed with the perceived economic potential of the industry (as yet unfulfilled) and are all being driven by the push factor of technology, which has created the impetus behind the progress of biotechnology in the last 30 years.

The regulation of biotechnology faces challenges as the biotechnology firm is expected to work on a global basis, whereas regulatory standards are often nationally based. This has been recognized by the OECD, who have established various working parties and attempted to encourage the development of internationally agreed protocols:

> The Working Group is also focusing on outreach activities, particularly through its information exchange mechanism, BioTrack Online. This mechanism includes information on regulatory developments in OECD Member countries, including details of laws, regulations and the contact points of the responsible ministries and agencies. It also has a database of field trials in OECD Member countries, as well as a database of products that have been commercialized. (OECD 2000, p. 8)

However regulation need not necessarily be a constraint on the biotech firm if it provides a clear and consistent environment in which to operate and highlights best practice standards that the biotech firm can use to promote itself to other regions. The UK, for example has recognized the importance of clear standards for biotechnology as a guide to firms, a means of encouraging development of critical mass but also the wider community, as indicated by the BioSafety Protocol developed in the late 1990s:

> The adoption of the Biosafety Protocol to the Convention on Biological Diversity in January 2000 is intended to lay the foundation for a global system for assessing the impact of genetically engineered organisms on biodiversity, and exchanging information through a Biosafety Clearing House. It also contains provisions to encourage capacity building in developing the environmental assessment of genetically engineered organisms. This experience shows that interactions between intergovernmental organizations and sharing of technical documents and expertise will avoid duplication of efforts and facilitate understanding of risk/safety assessment of products of biotechnology. (OECD 2000, p. 6)

Nevertheless, national regulation varies in its strictness which may severely constrict a firms' ability to operate. In some cases where a federated system operates, such as the US and Australia, complexity may be added by different regulation at state and national level. The recent example concerning restriction of stem cell availability in the USA but subsequent opening of these standards at state level, by California, for instance, provides such an example. In contrast, the Swiss people have recently endorsed stem cell research in a national referendum that saw support for such research from 66 per cent of the population – the first time that such a decision has been put to a national vote (BBC News, 2004). Other regions and states, for example Singapore, have deliberately relaxed regulation to attract new firms and researchers in the areas in an attempt to build critical mass of expertise and achieve global competitive advantage.

In the UK, the regulatory framework attempts to maintain a compromise between vigilance and loose control.

> While we agree that regulation should set clear, ethical limits beyond which researchers should not be allowed to go, public opinion in the UK seems broadly content with the difficult ethical balance struck in the regime here. We would therefore oppose any attempt to tighten regulation here. We are aware that the Government takes the same view, but we wish to underline the importance of continuing vigilance; the regulatory environment for biotechnology research in the UK is a real source of advantage and must not be undermined by developments at the European level. (House of Commons 2002–2003, column 14)

The key issue in regulation is achieving the correct balance of risk management, as summarized by Livingstone with respect to the Australian situation, 'Our challenge, therefore, particularly in the biotech area, is to devise sustainable risk management strategies, which will enable innovation to *proceed*, *with* appropriate precaution and regulation, and *without* alienating the broader population' (Livingstone 2002, p. 6, emphasis in original).

To the new biotech firm, understanding and negotiating the various regulatory requirements can be difficult and require significant expertise. Frequently more than one body or organization assumes responsibility for a single sector. As an example, key regulatory bodies in the USA, Canada and Australia by industry sector are given in Table 9.1.

In the UK, a range of strategic advisory bodies have been established to provide advice to the government relating to issues such as food safety and standards. These include the FSA, and the Human Genetics Commission, which works closely with the Department of Health. Applications for plant variety rights are handled through the European Community Plant Variety Office, while environmental remediation is covered by the Cartagena Protocol, the protocol on biosafety to the UN Convention on Biological Diversity (Secretariat of the Convention on Biological Diversity, 2005).

Governments are under increased pressure to consider more the food itself rather than the peripherals (packaging, residues, contaminants, labelling and processing methods). Significant efforts have been made to develop internationally agreed standards, especially by organizations such

Table 9.1   *Key regulatory bodies for the US, Canadian, and Australian*
           biotechnology industries, by sector

	Australia	Canada	US Agency
Human health and medical devices	Office of the Gene Technology Regulator (OGTR) National Occupational Health and Safety Commission (NOHSC) Therapeutic Goods Administration (TGA)	Health Canada Environment Canada Transport Canada Department of Foreign Affairs and International Trade Human Resources Development Canada	Federal Drug Administration (FDA) (Drug testing and approval)
Agriculture and crop bio-technology	National Registration Authority (NRA) Australia New Zealand Food Authority (ANZFA) Australia New Zealand Food Standards Council (ANZFSC)	Canadian Food Inspection Agency Health Canada Environment Canada Transport Canada Department of Foreign Affairs and International Trade Human Resources Development Canada	FDA (new foods and ingredients, safety and quality monitoring standards, labelling) USDA (new crop and variety controlled field trials) EPA (food residue tolerance levels, new food testing protocols) USDA and EPA (safety requirements and standards for pesticides, herbicides and genetically enhanced test crops)
Environment	EPA Australian Quarantine and Inspection Service (AQIS)	Environment Canada Health Canada Transport Canada Department of Foreign Affairs and International Trade Human Resources Development Canada	EPA USDA

Source: BIO 2004; BRAVO Canada 2003, Australian Government 2004

as the OECD, the joint food standards programme of the Food and Agriculture Organization and the World Health Organization, UNIDO, the International Organization for Epizootics; the Asia-Pacific Economic Cooperation (APEC) forum's Experts Group on Agricultural Technical Cooperation; the UN Environment Programme; and in the UN Biosafety Protocol. Discussion of genetically modified organisms also takes place within the Codex Alimentarius Commission. Such international agreements and regulation is intended to avoid trade disputes and improve public and consumer confidence in food safety (OECD 2000).

Environmental protection for the international movement of living modified organisms (LMOs) is provided by the Cartagena Protocol on Biosafety (2000). Cartegena was the first Protocol to the Convention on Biological Diversity (1992), and is the only international instrument dealing exclusively with LMOs. The Cartagena Protocol established a global mechanism for decision-making on imports and exports of LMOs, and provided a multilateral framework to assist decision-making and reduce risk for importing countries.

The most important regulatory issues affecting a health biotech firm are those relating to drug testing and approval. The lengthy process demanded has a significant impact on the ability of a firm to get its product onto the market, and is fraught with risk and high chance of failure. 'One-product' firms are especially vulnerable to collapse where the testing process fails, or may be subject to extreme share price swings if less than positive results arise from any stage of clinical trials. The long and complex process required by FDA is highlighted in Figure 9.1.

THE NATURE OF A GLOBAL INDUSTRY

The biotech firm needs to be particularly aware of the implications of its findings and how both researchers and the wider population will perceive end products. In cases where the end product represents a radical innovation, it is most important that researchers, the firm and the firms' executives are aware of likely implications so that they can commence a public awareness campaign at a sufficiently early stage. While there will always be early adopters of a technology, a radically new product may challenge deeply ingrained beliefs and therefore present a significant challenge to the wider community that may impede take-up and even cause rejection, and in the worse case scenario, result in extreme regulation or moratoria. Examples of this include GM crops, where even multinational companies such as Monsanto and Bayer failed to initiate basic public understanding of transgenic crops before their introduction, resulting in international bans and public protest.

In such circumstances, mass adoption of a technology is unlikely despite uptake by early adopters. For genetically modified crops, public awareness campaigns were probably commenced too late to be effective, with governments making decisions to ban crops because of public pressure and political damage. In such cases, demonstration of public, economic or social benefits is no longer effective because of the need to take a political risk-reducing decision.

The recent debate regarding genetically modified crops in Europe is an example. Despite UK approval, and pronouncement that certain crops were harmless, the European Union has continued to be deeply divided about approving growth of GM foodstuffs (Reuters 2004). The German Parliament recently passed laws strictly restricting cultivation of GM (genetically modified) plants, mainly as a result of political expediency.

These decisions have been made despite the increased world-wide uptake of GM crops and economic, social and environmental benefits, for example insecticide usage decreased by 67 per cent, yields increased by 10 per cent, generating income gains of US$500 per hectare from 1999 to 2000 in China. While in India, field trials demonstrated a 50 per cent reduction in insecticide spraying, and a 40 per cent increase in yields, equivalent to a 2.6 fold increase in income to US$200 per hectare (BIO 2004).

Experience by the multinational companies involved in development of GM crops suggest that new firms need to consider carefully the impact of the commercialization of their research, their products and public understanding of benefits and possible risks at an early stage. Partnering with larger firms is not necessarily a solution in itself in such cases. However, in the case of human health and particularly drug development, partnerships play an increasingly important role.

HUMAN HEALTH

Some areas of human health, such as radical tests for genetic defects in the foetus, parallel GMOs in being highly controversial. Others, such as development of new drugs for chronic diseases, are more readily accepted, and draw public support for more rapid introduction.

The regulation and approval of new drugs is a lengthy, time-consuming and expensive process. The FDA have attempted to streamline their approval process to ensure that new pharmaceuticals that show significant potency can be approved and introduced to the market more quickly, but the process is still lengthy and time-consuming, with a number of steps (Figure 9.1).

Since most biotech firms are not fully integrated, partnership with other

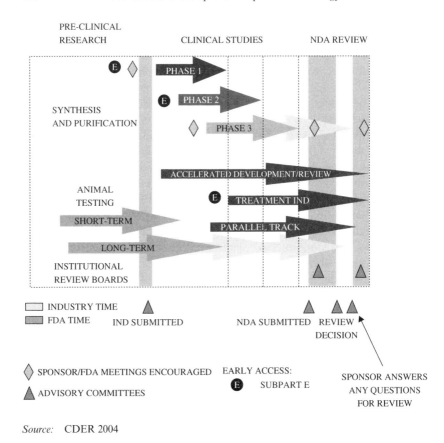

Source: CDER 2004

Figure 9.1   The new drug development process: steps from test tube to new drug application review

organizations is an essential part of the approval of a biopharmaceutical drug. The nature and complexity of the clinical trials process has led to the development of allied industries that focus on scale-up and manufacturing and clinical trials. Contract research organizations (CROs) now play an important part in the development of new drugs and are globally distributed. Increasingly new markets such as India specialise in providing CRO-type speciality to pharmaceutical companies, because of their capabilities in engineering, particularly bioprocessing.

India has a long tradition, not only in science and technology, but also in government support for biotechnology. The Department of Biotechnology was set up back in 1986. This can be viewed as an example of entrepreneurial opportunity recognition at the governmental level. This early support and consequent funding has led to success stories such as Biocon India with a successful IPO recently; Wockhardt which has launched three products – rDNA human insulin, a Hepatitis B vaccine and an Erythropoietin. There are also those success stories catering for the domestic market such as Bharat Serums which is manufacturing 125 million doses of oral polio vaccine (Grace 2004). With India a signatory to TRIPS from 2005, some doors will open, while others (largely in generics and products in which international IP has been ignored) will close.

From the perspective of Big Pharma, contracting testing out to other companies further reduces development risk and cost, removing the need for expensive expertise and infrastructure to be held in-house. In addition, regions seeking to build biotechnology capability seek to spread expertise from narrow drug development potential to a broader expertise.

CROs need to be able to conform to strict regulatory testing and checking standards. Where these are either not met, or are suspect, or incomplete or inaccurate records are kept, the result will be longer-term approval times and challenges to the acceptance of drugs on the global market. New players to the CRO market, such as India, have had problems in this respect, where companies were forced to withdraw generic drugs from the market due to improper documentation of bio-equivalence studies (Business line 2004). Instances such as this provide a major challenge to emerging whole industries rather than companies alone. However such examples are to date, isolated.

The drug regulatory process offers a significant challenge to the entrepreneurial biotech firm, given the complexity of testing stages and approvals required. The biotech firm therefore has more potential to attract significant partnerships with major pharmaceutical companies if it is able to offer a more integrated approach to drug development and testing. However, significant expertise is now required on the part of the biotech company as evidence suggests that partnering deals with major pharmaceutical companies are becoming fewer but larger (Burrill 2004).

One of the issues that remains high in public perception is that of ethics. This is briefly covered in the next section.

ETHICS

Ethics plays a key role in biotechnology, which challenges boundaries of technology and biological application. A Canadian report defined ethics as:

the activity of thinking about and deciding how people ought to act in their relationships with one another, or how human institutions and activities ought to be
organized. In another formulation, it is the application of moral values to
factual situations in order to determine how we ought to act in those situations.
(Canadian Biotechnology Strategy Task Force 1998)

The major factor driving ethics in the industry is socially acceptable use of
the biotechnology to 'save or improve lives, improve the quality and abundance of food, and protect the environment' (BIO 2004).

Scientific developments in biotechnology offer significant challenges to
thinking about life and the way that technology can assist life. At this stage,
ethical standards are largely industry imposed, with government regulation
covering particularly controversial or politically sensitive areas such as
cloning, genetically modified foods or use of biodiversity. It remains to be
seen whether self-regulation will continue or whether increased regulation
will be brought into force. Possibly the most controversial area is in the use
of stem cells from foetuses, and the increased ability for biotechnology to
offer genetic testing and design for the unborn child. Such controversies are
a long way from solution and likely to be inordinately effected by emotion
and political expediency rather than scientific discovery.

REFERENCES

BIO (2004), 'BIO Organisation Biotechnology Handbook', Available: www.bio.
org, Accessed: 25 November 2004.
Biotechnology Australia (2005), Available: http://www.biotechnology.gov.au/,
accessed 12 June 2005.
BRAVO Canada (2004), Available: www.strategis.ic.gc.ca, Accessed: 25 November
2004.
Burrill, G. (2004), Biotech 90: Into the Next Decade.
Business line (2004), 'The bio-equivalence trap, Express Healthcare Manage-
ment', Available: http://www.expresshealthcaremgmt.com/20041215/edit01.shtml,
Accessed: 30 December 2004.
Canadian Biotechnology Strategy Task Force (1998), *Making Ethically Acceptable
Policy Decisions: Challenges Facing the Federal Government*, Winter, p. 5.
CDER (Center for Drug Evaluation and Research) (2004), *Report to the Nation:
Improving Public Health through Human Drugs*, Washington, DC: Food and
Drug Administration, US Department of Health and Human Services.
Grace, C. (2004), *The Effect of Changing Intellectual Property on Pharmaceutical
Industry Prospects in India and China: Considerations for Access to Medicines*,
London: DFID Health Systems Resource Centre.
House of Commons Trade and Industry Committee (2003), 'UK biotechnology
industry', Twelfth Report of Session 2002–03, HC 87, published on 3 September
2003, London: Stationery Office.

Livingstone, C. (2002), 'Risk and regulation in biotechnology', keynote conference address to the Ninth Conference of the Global Harmonization Taskforce, New South Wales government, Sydney, June.

OECD (2000), *Report Of The Working Group on Harmonization of Regulatory Oversight in Biotechnology*, Paris: OECD.

Reuters (2004), Available: http://www.reuters.com, accessed 30 December 2004.

Secretariat of the Convention on Biological Diversity (2005), Available: http://www.biodiv.org/default.shtml, Accessed 1 September 2005.

10. The biotechnology value chain

INTRODUCTION

The Value Chain was first popularized by Michael Porter in his book *Competitive Advantage* (1985). The value chain refers to a network of activities, connected by linkages that are performed by an organization to design, produce, market, deliver and support its products and services. The value chain is essentially a framework for identifying the discrete but interconnected activities that make up a business and how those activities affect both the cost and the value delivered to buyers (Carr 2001). The value chains of organizations within an industry differ as a result of their particular strategy, their history and success at implementation (Porter 1985).

THE VALUE CHAIN

The organization's value chain is embedded in a larger stream of activities within the value system of the industry. The value system consists of supplier value chains (upstream) that deliver inputs to the organization. Downstream to the organization's value chain, products pass through the value chain of distribution channels and become part of the value chain of buyers. All activities in the value chain ultimately contribute to buyer value.

The basis for differentiation and competitive scope in an organization is its value chain and how this integrates within the value system. Some organizations may tailor their value chain for a particular industry segment that differentiates them from other organizations competing within the same industry. Other organizations may form relationships with another firm through joint venture, licensing or supply agreements that involves coordinating or sharing value chains as a means of broadening their effective scope.

The organization's value chain is composed of nine generic categories of activities that are linked together in a unique way (Porter 1985). The value chain shows how the activities of an organization are linked to each other and to the activities of suppliers, distribution channels and buyers. The value chain displays total value and consists of a number of value activities and the overall margin the organization achieves by delivering this value to customers. The value activities are the physical and technologically distinct activities performed by an organization. Every value activity involves some form of inputs, such as human resources and technology, to perform its function. Importantly, each value activity creates and uses information.

Value activities can be divided into two broad categories, primary activities and support activities. Primary activities are those involved in the physical creation of products and services. Primary activities can be divided into five generic categories including inbound logistics, operations, outbound logistics, marketing/sales and service. The support activities maintain and facilitate primary activities and each other, and include purchased inputs, human resources, technology and company-wide infrastructure.

Value activities are therefore the discrete building blocks of an organization's capability and competitive advantage. The way each value activity is performed will determine how well buyers' needs are fulfilled by the organization. Identifying the particular value activities in an organization involves the isolation of activities that are technologically and strategically distinct to the organization (Porter 1985).

The primary activities are further described below. The type of activity depends on the particular industry and the organization's strategy.

- *Inbound logistics* Activities associated with receiving, storing and disseminating imports to the product or service.
- *Operations* Activities assessing the inputs to the final product or service. These include development, assembly, testing, manufacturing, and packaging.
- *Outbound logistics* Activities associated with collecting, storing and distributing the product to buyers.
- *Marketing and sales* Activities associated with the sale of products to buyers, such as pricing, quoting, advertising, promotion and distribution.
- *Service* Activities associated with servicing or maintaining product parts, such as installation, training and technical support.

Specific activities or elements of the value chain are vital to a company's competitive advantage. Some organizations may focus on one or a few activities of the value chain to take advantage of their core competency. For example, Dell Computer Corporation focuses on the delivery component of the value chain where customers interface with the company through the Internet. They are able to deliver their build-to-order strategy by using the Internet to trigger activities among their suppliers and trading partners who make components and assemble computers – Dell does not manufacture computers it simply assembles made-to-order computers and delivers them direct to customers (Govindarajan and Gupta 2001). Dell redesigned the traditional value chain in the personal computer industry in a number of ways:

- It outsourced component manufacture, but performed assembly;
- It eliminated retailers and shipped directly to end-users;
- It took customized orders for hardware and software over the phone and via the Internet;
- It designed an integrated supply chain linking its suppliers to its assembly sites and order-intake system.

Managing the value chain of an organization as a system rather than separate parts is a key to gaining competitive advantage (Porter 1990). Reconfiguring the value chain by relocating, regrouping, reordering or eliminating activities as shown by the Dell example can improve the competitive position of the company.

To redesign or reconfigure the value chain a company must begin with a value chain analysis. Value chain analysis is concerned with identifying the input and output of economic resources found in a chain of value-adding processes or activities (Finger and Aronica 2001). The process involves reviewing the company's customer value proposition, determining how to transform the value customers receive, and redesigning the value chain architecture. Govindarajan and Gupta propose three principles to guide the redesign of a value chain (2001):

- The value chain's two central attributes must be redesigned – that includes the activities that will constitute the new value chain and the interfaces between the activities;
- The new value chain must create significant improvements in one or more of three areas: cost structure, asset investment and speed of responsiveness to external changes;
- The new value chain must enable the company to scale up its business model to ensure rapid growth in market share and expand its products or services.

One approach to value chain redesign is to shift from selling discrete products to supplying total solutions. For example, some computer companies, such as IBM, have reinvented their business system from selling computer hardware and software to supplying a total business solution. Companies that are successful in this approach are reinventing the concept of customer value by providing a one-stop-shop for clients.

THE CONCEPT OF VALUE

It is important to understand the concept of value and how this applies to business strategy and the customer model. Value is defined as the combination of benefits that are delivered to the customer less the total costs of acquiring the delivered benefits (Walters and Lancaster 2000). Value is therefore the difference between what a customer is willing to pay and the cost of producing the value. From a strategic perspective, an organization identifies the value drivers it is attempting to offer the target customers and the relevant activities involved in producing the value, together with the cost-drivers involved in the value-producing activities (Walters and Lancaster, 2000). The ability of a company to perform these specific activities more efficiently and economically than its competitors determines its competitive advantage (Brown 1997). How this value is to be delivered to customers is the company's value proposition. It has been suggested that the value proposition is the firm's single most important organizing principle because this is how the company strategically positions itself in the mind of the customer (Kothandaraman and Wilson 2001).

Slywotsky and Morrison (1997) proposed a totally different approach to value where the traditional value chain is totally reversed to concentrate on the customer. They suggest that a company first needs to focus on customer needs and priorities and then identify the channels through which the customer buys. The products and services best suited to the channels must be sourced or produced. Finally, the company determines the assets and core competencies required to meet customer needs (Slywotsky 1998). Reversing the value chain and placing the customer as the first link has also been described by Walters and Lancaster (2000) and Webb and Gile (2001).

THE VALUE NETWORK OR WEB

Porter's value chain model was originally presented as a linear chain functioning in a sequential order. This representation is suitable when describing or analysing activities within the organization. However, activities and exchanges between organizations, customers and suppliers may be complex and dynamic. Recent literature describes a value network or web as a more appropriate form (Finger and Aronica 2001; Allee 2000; Andrews and Hahn 1998). The value network has been defined as an interconnecting web of value-creating and value-adding processes.

A value network generates economic value through complex dynamic exchanges between one or more enterprises, customers, suppliers, strategic partners and the community (Allee 2000). The transactions involved go

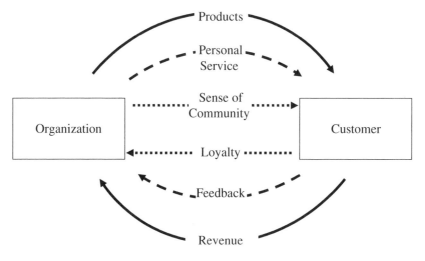

Source: Adapted from Allee 2000.

Figure 10.1   Value exchanges

beyond the exchange of goods, services and revenue. They also include the exchange of two additional currencies – knowledge value and intangible benefits (shown as broken lines in Figure 10.1). A mechanism or medium, such as computer systems and the Internet, support and facilitate the exchange of value.

Knowledge is becoming an important currency of exchange because knowledge can be exchanged for different types of value including money and other knowledge, an intangible benefit. Value networks can be complex. Mapping the value network requires that all three value exchanges, products/services, knowledge and other intangibles be identified and documented. Figure 10.2 shows an example of a value network between a typical pharmaceutical company and its internal and external value exchanges.

Finger and Aronica (2001) argue that the influence of the Internet and electronic commerce is transforming the linear value chain into a productive and efficient value network. They suggest that in the digital economy that is, the modern economy where information technology is a key driver, the once large, rigid value chains must become lightweight, fine grained and adaptive. The result is a web of connections that drive the supply chains, demand chains and business processes that represent core competencies of an enterprise. These new value networks or webs are dynamic and customer-driven.

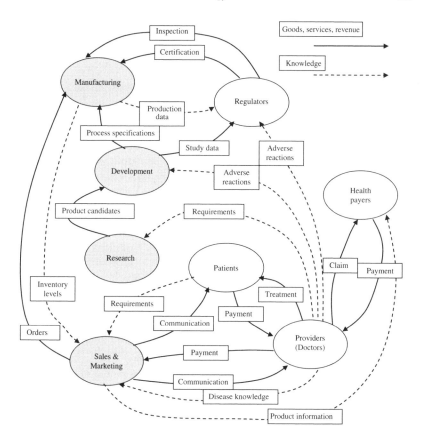

Source: Adapted from Alee 2000.

Figure 10.2 Value network analysis diagram

Andrews and Hahn (1998) support the view that value chains are being transformed into value webs. They suggest that there are two driving forces reshaping value chains:

1. Continuous changes in the roles of value chain members; and
2. Customer/consumer preference for personal customization and quick gratification.

The value network also provides the infrastructure to manage and facilitate the exchange of information/knowledge, critical to the business, in real-time and respond appropriately to the needs of the customer.

Table 10.1   The shift from value chain to value web

Criteria	Value chain	Value web
Customer	Focus on customers	Focus on end-consumers
Environment	Static/stable	Dynamic/changing
Scope	Domestic or multinational	Global
Focus	Industry-specific	Cross-industry
Value creation	Independent	Partner-based
Relationship type	Business relation	Strong partnering
Infrastructure integration	Limited integration	Full integration
Infrastructure thrust	Cost-driven	Value-driven
Infrastructure currency	Reliability and accuracy	Speed and synchronization
Process focus	Improves own processes	Improves joint processes
Profit focus	Increase own profits	Increase profits of all
Cost focus	Considers own cost	Considers total cost
Knowledge leverage	Within the enterprise	Across all nodes
Knowledge approach	Hoarding	Sharing
Resource approach	Guarding	Sharing

Source: Adapted from Andrews and Hahn 1998

The more sustainable organizations are likely to be those that are able to continuously collect, archive, retrieve, create, share and leverage information that can be translated into knowledge and wisdom for real-time decision-making (Andrews and Hahn 1998). Table 10.1 compares the major characteristics of a value chain and a value web.

THE BIOTECHNOLOGY VALUE CHAIN

The pharmaceutical industry, which is closely related to the biotechnology industry, began in the early nineteenth century when scientists, after realizing the therapeutic effects of herbs and plants, began to extract and systematically isolate specific active ingredients (Champion 2001). These active ingredients were then tested for efficacy and sold as therapeutic pills. The steps undertaken to develop these early pharmaceutical products involved firstly research, then testing and finally, delivery. These three steps defined the industry's value chain for the next 100 years (Champion 2001).

During the 1960s after the discovery of DNA, it was realized that certain diseases are closely linked to genetic constituents. Specific genes influenced changes in proteins that were responsible for disease. The focus moved towards basic research, which was identified as an important function in the discovery of new compounds. During this stage, significant R&D resources were directed towards the identification of genes associated with the disease and the proteins (targets) these genes produced.

Models or schematic representations of the pharmaceutical value chain have been set out in numerous publications (BIO 2001; Ernst & Young 2000; Champion 2001; Myers and Baker 2001; Granberg and Stankiewicz 2002; Shillingford and Vose 2001). These models represent the pharmaceutical value chain as a series of activities or steps (elements, functions, phases). A schematic model of the expanded value chain for the pharmaceutical industry is outlined in Figure 10.3. This value chain is typical of the modern pharmaceutical company for example, Millennium (Champion 2001), that covers the full spectrum of activities, from early discovery research to new drug approval, through to scale-up manufacturing and marketing of the final product. The clinical phase steps may be broken further into sub-steps incorporating phase I, phase II and phase III clinical trials and registration of the product with bodies such as the FDA in the USA. The regulatory activities of the pharmaceutical value chain model outlined in Figure 10.3 may not necessarily follow a sequential pattern, however all steps need to be completed before market launch. Some of the activities may overlap or occur in parallel. For example preclinical development and testing may continue as a drug candidate enters the clinical trial phase. The models in the literature may differ slightly but have a basic similarity given the highly regulated nature of the overall pharmaceutical process.

The major activities of the pharmaceutical value chain are described in more detail below. The first step in the chain is the identification of genes that are involved in the particular disease. The next step is to identify disease proteins or potential targets that the different genes produce in different parts of the body. The proteins or targets are those that cause malfunctions in cells and subsequently result in disease. Following this, small molecules able to attach to the target protein are identified. This research step is known as lead identification. These small molecules have the potential to prevent the disease from occurring by blocking specific active sites or receptors when they attach to the proteins. The small molecules become the leads, which are optimized and then enter the testing stage. Pre-clinical testing is initially performed on an animal model, which includes toxicity testing of the lead compound. This is followed by clinical trials in human populations.

Clinical trials are segmented into three discrete phases. In a Phase I trial, a new drug is tested on humans for the first time. The objective of this phase is to assess the safety of the drug in relation to dosage and side effects.

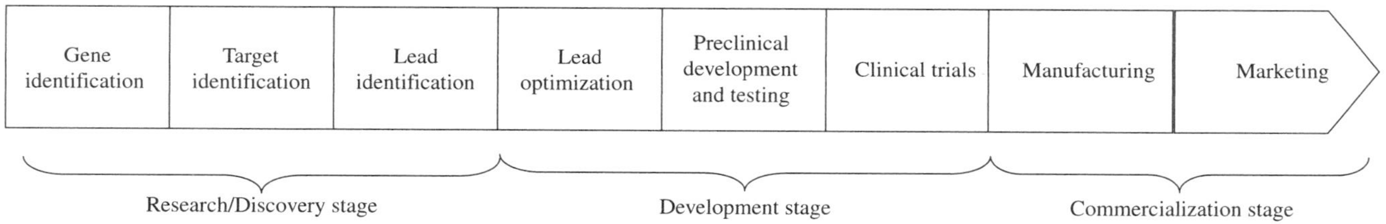

Source: Adapted from Champion 2001.

Figure 10.3   The expanded value chain of the pharmaceutical industry

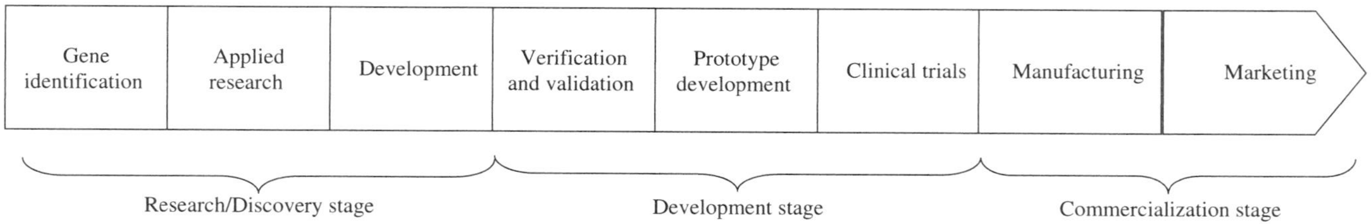

Source: Kapeleris et al. 2004

Figure 10.4   The elements of the biotechnology firm's value chain

Phase I trials are relatively inexpensive and involve 20–80 clinical trial subjects. Phase II trials provide preliminary information on the efficacy of the drug, including how well the drug works and additional safety information. This phase requires approximately 100–300 clinical trial subjects to conduct the trial. Phase III is large and extensive requiring 1000 to 5000 clinical trial subjects with an average cost of US$7500 per subject (Stewart et al. 2001). These trials test and compare the new drug with combinations of drugs or with an existing standard therapy. Phase III trials take a significant amount of time to complete and at a considerable cost to the company.

The last two components of the pharmaceutical chain include manufacturing and marketing, and together make up the delivery phase. These activities do not occur in large scale until the regulatory body, such as the FDA, has approved the sales and use of the new drug. During the late stage clinical trials, preparations for manufacturing and pre-marketing of the product usually occur in parallel. Once registration is obtained for the new drug, large-scale manufacturing of the compounds occurs, at the same time ensuring costs are kept to a minimum. The products are then marketed to doctors and patients.

Resource limitations and availability of funding shape the value chains of pharmaceutical organizations. The pharmaceutical company will focus on its area of expertise and then partner with another pharmaceutical company to further develop its products or commercialize any promising drug candidates (Powell et al. 1996). An example of this involves gene identification, which is not performed by all pharmaceutical companies. This activity is performed by specialist biopharmaceutical companies that focus their activities in this area, using differentiated and patented procedures for screening and selection. Revenue for these firms is obtained through licensing out the lead molecules to a larger pharmaceutical company or a company that may specialize in another set of activities downstream in the value chain (Woicheshyn and Hartel 1996). The small biopharmaceutical company does not usually have the required capital (significant cost associated with Phase III clinical trials) to take a pharmaceutical product to market through its own efforts. It must partner with a larger, established pharmaceutical company to achieve the objective.

Figure 10.4 outlines a biotechnology organizational value chain model consisting of eight major elements extracted from the aggregation of the value chains from the multiple case studies. The value chain has been outlined in a linear format to depict the major generic activities or elements where value is created within a typical biotechnology firm.

The first step in the chain is the generation of a novel idea through some form of basic research. This process usually occurs at a university or

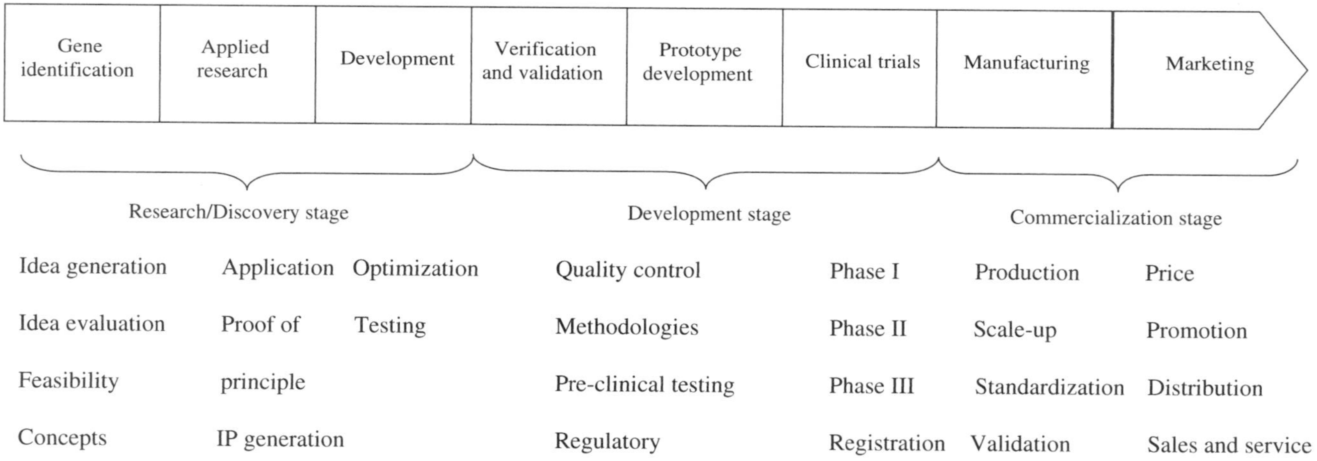

Gene identification	Applied research	Development	Verification and validation	Prototype development	Clinical trials	Manufacturing	Marketing
Research/Discovery stage			Development stage			Commercialization stage	
Idea generation	Application	Optimization	Quality control		Phase I	Production	Price
Idea evaluation	Proof of	Testing	Methodologies		Phase II	Scale-up	Promotion
Feasibility	principle		Pre-clinical testing		Phase III	Standardization	Distribution
Concepts	IP generation		Regulatory		Registration	Validation	Sales and service

Source: Kapeleris et al. 2004

Figure 10.5   The value chain of the biotechnology firm

research organization. The next step is to apply this research to some form of practical use. This involves a proof-of-concept stage. Completing the proof-of-concept demonstrates some form of utility, and if the concept is novel then a patent can be filed to protect the discovery. Following this, the concept will enter development where further optimization takes place. This may involve the development of a product that is based on the concept.

Once the product is optimized and quality controlled through verification and validation processes, a working prototype is developed. The prototype is tested or trialled in a clinical setting to ensure it performs as required. Clinical trialling ensures the safety and efficacy of the product. Once clinical testing is completed further modification may be required. When the modifications are completed the product is ready for commercialization. This begins with the scale-up manufacturing of the product and then final delivery to the customer through the marketing function.

The generic model can be further broken down in to specific sub-activities that are performed by different organizations in the biotechnology industry as Figure 10.5 demonstrates. These sub-activities demonstrate the diversity of tasks performed in the biotechnology industry. This revised generic model takes into consideration standard sub-tasks that were identified in the case studies.

The drug discovery, design, development and commercialization activities that are typical of the pharmaceutical value chain could also fit within the biotechnology value chain model, outlined in Figure 10.5, if the sub-tasks are also identified.

CONCLUSION

The pharmaceutical industry is closely linked to the biotechnology industry. Many small to medium biopharmaceutical companies are involved in drug discovery that feeds the larger pharmaceutical companies. Biotechnology firms resemble major pharmaceutical companies in the length, cost and regulation of their product life cycles. Most biotechnology companies are often small, entrepreneurial firms with limited resources that focus their activities on the science or technology platform that drives their existence. These companies tend to focus less on the manufacturing, marketing and distribution of final products. Companies that manufacture, market and distribute their own research and development are able to create value across the spectrum of the value chain.

REFERENCES

Allee, V. (2000), 'Reconfiguring the value network', *Journal of Business Strategy*, July/August, 36–9.

Andrews, P.P. and J. Hahn (1998), 'Transforming the value chain into value webs', *Strategy and Leadership*, **26** (3), 6–11.

BIO (2001), 'Editor's and reporter's guide to biotechnology', report by the Biotechnology Industry Organization, USA.

Brown, L. (1997), *Competitive Marketing Strategy*, Melbourne: Nelson.

Carr, D. (2001), 'Forging value chains for the 21st century', *Internet World*, **47** (14), 27–34.

Champion, D. (2001), 'Mastering the value chain', *Harvard Business Review*, June, 109–15.

Ernst and Young (2000), Convergence: The biotechnology Industry Report.

Finger, P. and R. Aronica (2001), 'Value chain optimization: the new way of competing', *Supply Chain Management Review*, September/October.

Govindarajan, G. and A. Gupta (2001), 'Strategic innovation: a conceptual road map', *Business Horizons*, **44** (4), 3–13.

Granberg, A. and R. Stankiewicz (2002), 'Biotechnology and the transformation of the pharmaceutical value chain and innovation system', unpublished work, Research Policy Institute, Lund University.

Kapeleris, J., D. Hine and R. Barnard (2004), 'Towards definition of the global biotechnology value chain using cases from Australian biotechnology SMEs', *International Journal of Globalisation and Small Business*, **1** (1), 79.

Kothandaraman, P. and D.T. Wilson (2001), 'The future of competition: value-creating networks', *Industrial Marketing Management*, **30**, 379–81.

Myers, S. and A. Baker (2001), 'Drug discovery – an operating model for a new era', *Nature Biotechnology*, **19**, 727–30.

Porter, M.E. (1985), *Competitive Advantage: Creating and Sustaining Superior Performance*, New York: Free Press.

Porter, M.E. (1990), *Competitive Advantage of Nations*, London: Macmillan Press.

Powell, W.W., K.W. Koput and L. Smith-Doerr (1996), 'Interorganizational collaboration and the locus of innovation: networks of learning in biotechnology', *Administrative Science Quarterly*, **41** (1), 116–45.

Shillingford, C.A. and C.W. Vose (2001), 'Effective decision-making: progressing compounds through clinical development', *Drug Discovery Today*, **6** (18), 941–6.

Slywotsky, A.J. (1998), 'The profit zone: managing the value chain to create sustained profit growth', *Strategy and Leadership*, **26** (3), 12–16.

Slywotsky, A.J. and D.J. Morrison (1997), *The Profit Zone*, New York: Wiley.

Stewart, J.J., P.N. Allison and R.S. Johnson (2001), 'Putting a price on biotechnology', *Nature Biotechnology*, **19**, 813–17.

Walters, D. and G. Lancaster (2000), 'Implementing value strategy through the value chain', *Management Decision*, **38** (3), 160–78.

Webb, J. and C. Gile (2001), 'Reversing the value chain', *Journal of Business Strategy*, March/April, 13–17.

Woicheshyn, J. and D. Hartel (1996), 'Strategies and performance of Canadian biotechnology firms: an empirical investigation', *Technovation*, **16** (5), 231–43.

11. Biotechnology industry and firm structures

INTRODUCTION

This chapter discusses the strong links between the structure of individual organizations and the entire industry; highlighting how both are influenced by scientific and technical developments occurring within the industry, as well as those from external industries that diffuse across to impact on biotechnology (such as bioinformatics) (Deliottte, Touche Tohmatsu 2003). The analysis considers size, growth and firm structures; the structure of entrepreneurial biotechnology firms; organizational structure and strategy; industry structures (including a brief discussion of Schumpeter, Weber, Ricardo and Malthus); technical impacts on structure; and industry trends and knowledge. These developments are important considerations for the entrepreneurial biotechnology firm as it plots its future path.

While this is a global industry, there are clear differences in industry structure at the national and regional levels. The role of universities, medical and research institutions is central to the industry in Australia, the UK and in Germany. This is particularly so given the embryonic and early developmental stages of these industries. This does not, however, negate the fundamentally influential role of the Big Pharmas in the development of each of these industries. Comparative national industry cases will serve to explore these structural variances in this chapter.

THE STRUCTURE OF THE NBF

As with most small firms, the NBF is as much a product of its environment as it is a strategically developed organization. Where the NBF differs from more typical small firms is in its structure which stems largely from its funding sources and its origins. Most NBFs are spinouts, spin-offs or developments from research projects and programmes, based largely around the intellectual property they possess. The NBF is almost always a product of its institutional environment. This differs dramatically from the traditional small business which emanates as an independent enterprise unsupported by large firms or institutions. The old adage of the requirement of the 3Fs to establish a small firm – friends, family and fools does not hold true for the NBF. It is from institutions they emerge and from institutions they seek funding and support. Friends, family, second mortgages, bank overdrafts and loans rarely come into the equation. From these origins it is not surprising that the organizational structure that emerges, and in fact the entire industry structure is somewhat unique.

Industry Types Where Entrepreneurs are Found

There are three industry types becoming increasingly evident in industrialized economies:

- *Established industries*   these are the traditional industries established in the industrial era which are capital-intensive. Competitiveness is achieved for individual firms by achieving through-put of product, the reduction of long-run average costs through volume output. They include industries such as winemaking, beer brewing and production, agrichemistry, and pharmaceuticals.
- *Emerging industries*   These are industries with a relatively short history, though a sufficient track record of growth and market success has been achieved. They are usually offshoots of established industries where firms have sought to establish new markets with new or changed products. This would include many areas of biotechnology, such as cell therapies, gene therapies, cell culturing and many areas of diagnostics.
- *New industries*   These have their origins either in the emerging industries or directly from the established industries. The differences between these and the emerging industries is that they do not yet possess a track record of success in product development and market penetration. e.g. nanomics, proteomics, metabolomics.

The biotechnology industry could still rightfully be considered to be an emerging industry. In most countries it is still dominated by large numbers of small firms which themselves are dependent upon public expenditure for support and survival. The common characteristics of new market and industries are listed below. The only feature of this list which appears not to fit the profile of the biotechnology industry is the existence of few standards. It would seem at first glance that this is an industry that is riddled with standards such as ethical standards, good laboratory practice, good manufacturing practice, gene technology regulation and stringent clinical trial processes. However in the development of the industry and its products there is no dominant design in the research and development processes which will tend to create the industry standards that all must follow. This approach is much more clearly evident in the IT industry, particularly in hardware design and in many areas of software design. This standardized approach can be very limiting to innovation.

Common characteristics of new markets and industries

- Few direct competitors
- Co-operation and collaboration between firms
- Room for smaller firms to operate
- Few standards in place
- Little if any price competition
- Substantial room for growth of firms and market
- Innovative climate – especially product innovation
- Rapid change in the environment
- High birth and death rates of firms
- Economies of scope more important than economies of scale
- Technology often plays a significant role
- Information flows are open
- High degree of product differentiation.

SCALE, SCOPE AND COMPETITIVENESS

The small business literature has given a great deal of attention to small firm competitiveness in growing industries, which, as indicated by the data above, provides an important, but only partial explanation of the growth in small business employment. Further attention needs to be given to dynamics within the manufacturing sector, where the shift to small firm employment is also important in understanding small firm employment growth. This chapter seeks to explain those dynamics with reference to the strategies of both small and large firms. The emphasis of small firms on scope rather than scale has given them a competitive edge in many sectors including biotechnology. At the same time, large firms in the pharmaceutical industry have been downsizing as they merge. However these large firms might in future be expected to pick up on the advantages of scope gained by the NBFs in their pursuit of competitive strategies. In this way the study of industry must assume a cyclic pattern over time, rather than a linear pattern in which one business model or form will always dominate.

Chandler's own definitions of economies of scale and of scope are referred to in this chapter:

> Economies of scale may be defined initially as those that result when the increased size of a single operating unit producing or distributing a single product reduces the unit cost of production or distribution.
> Economies of joint production or distribution are those resulting from the use of processes within a single operating unit to produce or distribute more than one product (I use the increasingly popular term economies of scope). (Chandler 1990, p. 17)

Chandler tends almost always to combine the two concepts, whereas other economists tend to separate the two. In actuality while both can occur simultaneously, firms will focus on either standardization in pursuit of scale, or diversity and flexibility in pursuit of scope. It is difficult for one firm to achieve both scale and scope given resource, management, cultural, structural and skills limitations.

Since 1975, modern capitalist economies have experienced significant change in the structure of industry, the nature of production and the organization of work. Technological developments have transformed production by expanding the possibilities for flexible and decentralized production techniques. Technological advances, have altered the production process and facilitated production of specialized products from generalized machinery, particularly in high throughput screening, nuclear magnetic resonance (NMR), polymerase chain reaction, advances in mass-spectrometry and crystallography. The integration of design and production functions (rather than their separation as in Fordism) has been possible because of flexibility in production arising from new technologies (Sabel 1995). This has had a major impact on many processes in biotechnology where the focus has shifted for pure science to production using these technological advances and de-emphasizing the scientific skills of researchers, technicians and scientists.

Economies of scale are thought to have been eroded by the advent of the more recent technologies, particularly those which are generic and commercial-off-the-shelf (COTS), as previously achieved cost advantages from scale and volume are minimized by dramatically reduced input costs. Even if an NBF can't afford to purchase a major item of equipment, they can lease time or outsource to the equipment provider, thereby achieving their competitiveness and still avoiding the need for scale. In fact economies of scale are about scale volume of production. This is a concept which is yet to come into the reckoning of most NBFs, as they attempt to develop early-stage products. For them the final consumer market is a faint glimmer viewed from a great distance.

Aligned to the technological developments has been the well documented emergence of a 'knowledge economy' in which the intellectual and human capital required for the full and effective utilization of the new, usually information-based technologies, places people at the centre of the production process, rather than as extensions of the technology.

While the assumption is made that small is beautiful, this is a false attribution. While smaller firms possess the features of flexibility, responsiveness, innovativeness, product and process innovation, opportunity recognition, and niche satisfaction; these are more essentially aspects of economies of scope possessed by small firms rather than essential features of small firms. Economies of scope are then not the exclusive domain of small firms. Small firms to this point have enjoyed the advantages made available to them by the knowledge economy. However there is mounting evidence, that the more competitive larger firms have begun seeing the light and altering their corporate strategy in line with economies of scope strategies.

The false attribution is that small firms are competitive because they are small. In actuality the small firms are competitive because they are employing economies of scope strategies naturally. This concept is displayed in Figure 11.1.

If there is a shift in focus amongst large firms, from scale to scope, then the advantages gained by small firms will be eroded and their competitive position will decline in the future. Once Big Pharma work through their 'scale crisis' to the 'scope solution', then their competitiveness can be regained. Potentially, NBFs will be at the whim of the strategies of these large firms again. This is where a shakeout is likely to occur in the biotechnology industry.

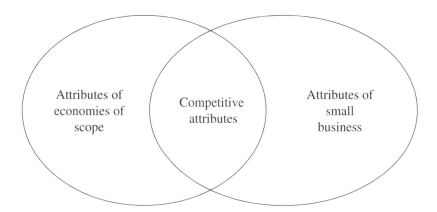

Figure 11.1   The false attribution dilemma

THE BIG FEAR – SHAKEOUT IN THE INDUSTRY

There was a proliferation of IPOs in 2003 and 2004 as the biotechnology industry recovered from the 2000 tech crash. In Australia alone were 18 IPOs in 2004 raising over $130 million (EG Capital 2004). There are now 105 listed biotech firms in Australia. There is a danger in this recovery however, of the emergence of a boom/bust cycle for the industry. This is alluded to in the previous chapter. However, the major problem here is that these cycles do not coincide with any other cycles in the industry. While the 2000 crash slowed the growth prospects of NBFs around the world, what is looming in the near future is potentially more cathartic.

The potential exists, following this latest surge, for a blow-out and consequent shakeout in the industry. Take by way of example, Neuren, a New Zealand biotechnology firm whose CEO made a presentation at a recent investment expo. The CEO was proud of the fact that they had seven research projects running, with partners from around the world, and the product pipeline would end in two distinct markets. The science was good, but the prospects for an investor were less definable. Representatives from funds throughout Asia, including Hong Kong and Singapore, were in attendance. Discussions indicated there would be a concern to invest in that company not knowing which direction they were going and also the extensive resources that would be required for a small company to run seven research projects concurrently. They may simply run out of money because the burn rate on each of the projects will be high, and the administrative costs of coordinating all these projects with international partners would be an added burden. A more prudent path would be to undertake a more stringent decision-making process early in the research to cull from the outset, and then to maintain a monitor of feasibility for each project and be able to make the hard decision to kill off projects where needed. Their only alternative to that is to spin-out companies, yet this isn't very viable as the spin-out will need financial support for its early existence. Neuren is not alone in its current status. With so many smaller biotechnology firms facing similar dilemmas, it seems only a change in industry structure will lead to a more sustainable future for many of these current firms. Just how this structural change will manifest itself can only be speculated at.

Mergers may be undertaken to achieve economies of scale, so that cost efficiencies are achieved through expansion of plant facilities while maintaining a concentration on a narrow range of products. This would tend to mean that close or direct competitors would be likely to merge. Integration will be likely to be both horizontal and vertical, while remaining in a single sub-sector.

However in this industry the preferred merger design will be aimed at achieving economies of scope, so that efficiencies can be gained from broadening the product range through the merger and utilizing existing lab space, staff and equipment more fully, while extending the knowledge base of the company. This will mean that the likely merger partners will not be close competitors; however, they will see complementarity in their research and their product base. Integration will be vertical and more broadly horizontal, potentially being between companies from different sub-sectors of the industry.

This potential new environment facing the biotechnology industry, based around research networks and three way alliances between government, firms and research institutions needs to be considered on its merits. To do this a framework for analysis requires construction to assess alliances and networks. Some possible well-known and accepted analytical perspectives include:

- Resource-based view
- Institutional theory
- Social exchange theory
- Absorptive capacity
- Transaction cost economics
- Evolutionary economics
- Organizational learning and knowledge transfer.

We have explored absorptive capacity and organizational learning and knowledge transfer in some depth in Chapter 6. A number of the remaining perspectives are based upon assumptions of a commercial exchange environment largely between firms. Research networks take an alternative form. This means that while the firm-based approach is not entirely irrelevant, we will need to 'pick the eyes out of' these perspectives.

The most relevant of the perspectives to research networks which involve university, government and private partnerships appear to be:

- Institutional theory
- Social exchange theory and
- Organizational learning and knowledge transfer.

The reasons for this level of relevance is that institutional theory already incorporates public/private partnerships while social exchange theory is not limited by the assumption of a commercial transaction exchange with a competitiveness agenda attached. In fact, social exchange theory aligns quite closely with other literature around networks, namely communities of practice.

Institutional theory is a highly relevant perspective in any study of research networks, given the need for involvement of institutions of some form, namely research/educational institutions government and major private sector players. Biotechnology is an industry dominated by institutional players and small firms. It should be recalled that institutional theory has led into two other distinct areas of enquiry related to innovation – triple helix and systems of innovation (Lundvall 2002), which will be discussed briefly later in this chapter.

Social exchange theory, upon which social network theory is founded, views the relationships, the links in a network rather than the players, the nodes. For example, Ahuja's (2000) work employing social network theory in the analysis of networks found that the number of direct and indirect ties a firm has affects its capacity to innovate and the effect of indirect ties on innovation is moderated by direct ties. Biotechnology is a highly networked industry, with communities of practice an important professional and social feature to the industry. Linked to work on social networks is social capital. According to Ireland et al. (2002, p. 429), 'although social capital is a public good or organizational resource, it is built through networks of personal relationships'. Hence the social relationships, social exchange, social networks and communities of practice are all vital aspects to analysing the processes involved in establishing and maintaining research networks.

Social exchange is not an end in itself. It is a means to knowledge generation and knowledge sharing, particularly in research networks in which competitiveness in the commercial sense is not a central driver. Organizational learning and more particularly knowledge transfer provides the third major line of enquiry in this study. The sharing of knowledge, both codified and tacit (Polanyi 1966; Nonaka and Takeuchi 1994) is a major driver of the collaboration process. Collaboration then brings with it economies of scope associated with sharing and accessing complementary resources (Gulati et al. 2000; Doh 2000). Different but complementary resources make it possible to gain economies of scope, create synergies and develop new resources and subsequent skills (Hitt et al. 2001).

The usual problems and pitfalls associated with alliances such as opportunism born out of self-interest (Williamson 1985), lack of trust (Sivadas and Dwyer 2000), power differentials, lack of commitment and problems with disclosure are mitigated in a network where the pressure for commercial competitiveness is lessened. For example, trust is less of a concern because the commercial exchange imperative is largely eliminated and communities of practice ensure or at least support knowledge sharing and minimize the self-interest opportunism of concern to transaction cost economics (TCE) and the resource based view of the firm (RBV), and dynamic capabilities scholars and proponents.

Organizational learning and learning in alliances need to be understood in depth. It is also important to explore the cross-disciplinary and inter-disciplinary nature of many biotechnology collaborations in this learning context. The role of the network here is critically important. According to Ireland et al. (2002, p. 413) 'partners learn from each other only when their knowledge bases are at least somewhat similar'. Such a sentiment relates to the communities of practice which will play an important role in the research network. However this perspective creates its own limitations when exploring cross and inter-disciplinary alliances, as it relies on the precondition of knowledge symmetry. Whereas knowledge asymmetry, reliance on differential knowledge bases, which would be more apparent in some research alliances may be more desirable – at least in terms of the scientific and technical knowledge – while other areas such as management and networking approaches should be more closely aligned.

'Understanding how alliances are formed and successfully managed requires the study of processes, including those designed and used to effectively manage alliances' (Barringer and Harrison 2000).

ALLIANCES AND NETWORKS LEADING INTO THE TRIPLE HELIX

The three strands of the triple helix is constituted of government, research and academic institutions and private sector. In Etkowitz and Leydesdorff's (1999) 'triple helix', governments, local, provincial and national, educational institutions, research institutes, private sector players and other community and representative bodies are all represented as active players in the biotechnology network. It should be recalled that institutional theory has led to two other distinct areas of enquiry related to innovation – triple helix and systems of innovation (Lundvall 2002).

Technological change has long been related to the generation of long-term economic growth (Landau and Rosenberg 1986; Tuma 1987). Studies from various countries have exposed this relationship. In the USA, 50 per cent of economic growth has derived from its technological advancement in the past 50 years. The figures are even stronger for France at 76 per cent, West Germany at 78 per cent, the UK at 73 per cent and Japan at 55 per cent (Mitchell 1999). These success stories from such developed nations have encouraged other developing nations to employ technology-based growth as one of the major strategies in achieving competitiveness, simultaneously improving living standards and levels of national prosperity (Jegathesan et al. 1997).

Governments of many countries have been seeking strategies to enhance as well as to manage their technological capabilities. The concept of a national innovation system (NIS) that emerged in the middle of the 1980s has been taken up as an appropriate means of achieving this end. It is specifically concerned with the creation, diffusion and utilization of innovation through interaction among many actors within national boundaries (Freeman 1987; Lundvall 1995).

The reason why networks and alliances are so important in the biotechnology industry, and why economies of scope are sought through the triple helix and innovation systems approaches, is exemplified in Figure 11.2 showing the benefit flows for the stakeholders of a certified good manufacturing practice (cGMP) facility which incorporates a business incubator. Confidentiality does not permit a full identification of the project, though this does not detract from the lessons to be learned in this case.

TRENDS IN THE STRUCTURE OF THE BIOTECHNOLOGY INDUSTRY

It is difficult to ignore the reality that large biotechnology companies (Big Biotech) are continuing to usurp the pre-eminent role of Big Pharma. Evidence of this is in the deals being undertaken between the NBFs and these newly emerging Big Biotech companies. It is increasingly likely that the NBFs will license out to the larger biotech companies as they see the R&D strengths of these companies extending their own core competencies, while Big Pharma increasingly concentrate on their distribution and marketing channels. This is exemplified by the Peptech's relationship with Abbott Laboratories and Centacor.

All firms and groups of firms wax and wane in their market influence. Given the current status of the value chain and its slow but inexorable movement towards the market, large firms will remain important as their scale in areas such as logistics, distribution, legal and management resources, as well as manufacturing capacity remain important. As industries mature, as discussed in the previous chapter, despite the pace of change creating its own destructive power, concentration levels increase. As the role of the Big Pharma wane, Big Biotech is waxing.

For NBFs, this creates opportunities for commercialization, either through development deals and associated milestones and eventually royalty payments, or through alliances, or mergers, acquisitions, or institutional involvement in IPOs. Despite the importance of NBFs to the development of the biotechnology industry, it is not the NBF which create, consciously, the industry strategies which lead the development of the

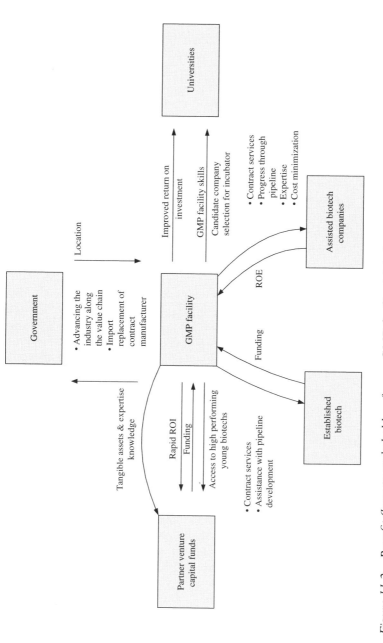

Figure 11.2   Benefit flows to stakeholders from a GMP facility initiative

industry. This is why concepts such as the triple helix are important, as industry strategies emerge from institutional involvement by long-term players with long-term visions. Small firms particularly are notoriously operational in their vision and find it difficult to strategize until they achieve a critical mass through a product range and resource base.

Strategy in industries dominated by small firms is at best emergent, at worst haphazard and random. This dilemma is apparent in many countries with immature biotechnology industries. The Danish biotechnology industry is a case in point. There are no flagship companies providing an innovative benchmark for other firms to follow. In 2003 there were 181 core biotechnology companies in Denmark, with an additional 86 with significant biotechnology activities. This is a dramatic increase over the 61 core biotechnology companies which existed in 1997 (Bloch 2004). This rapid growth, which now ranks Denmark fifth in Europe in terms of number of biotechnology companies, and the immature nature of the industry in Denmark means that while large enterprises and pharmaceutical companies are operating in the industry, there are no significant core biotechnology companies taking a lead in the industry (Ernst and Young 2003). Of those 181 core biotechnology companies, 149 had been founded since 1997 (Bloch 2004).

The immaturity of companies, coupled with the institutional basis of basic research in the industry, means it is institutions which take the lead in establishing the direction of the industry at a national level. The nominated institution generally is government. In fact this is an interesting dilemma for this institutional player, as much of the evidence in biotechnology is pointing towards the impact of metagovernance. Metagovernance has a number of connotations, however in an industry with a global stance like biotechnology, it tends to be the international players which tend to lead technological developments, as well as most strongly influence industry standards across countries. In pharmaceuticals it will be Big Pharma such as Roche, Pfizer, GSK, Novartis, Sanofi-Aventis who, while not concentrating on mergers and acquisitions, will lobby and influence national governments to support and maintain their industry regime, such as the patent system, pharmaceutical benefits schemes, research funding programmes, regional development policies and even curriculum and research programmes in universities. The more global the industry the more this is feasible to achieve and the less influence national and provincial governments will have over the broader industry agenda.

As a result there is a negative impact on the importance of national governments through metagovernance, led by innovative large and medium companies and international institutions such as WIPO, WTO,

Table 11.1   Competing influences on the role of government in the biotechnology industry

Influences which reduce government power in the biotechnology industry	Influences which increase government power in the biotechnology industry
Globalization and internationalization of trade and markets	Emergent strategies
Focus on the US market by many biotechnology companies	Institutional nature of research
Metagovernance usurping the role of national governments	Limited corporate leadership opportunities
The pace of innovation and change	Lack of critical mass
The value chain bubble moving towards the market or stopping	Reliance on public expenditure
	Importance of higher education
	Importance of regions and industrial districts to the development of the industry in many countries

and international industry associations, which take the lead in the global direction of the industry. Table 11.1 explores some of these competing influences.

COMPARING THE BIOTECHNOLOGY INDUSTRIES AND ENVIRONMENT IN AUSTRALIA AND THE UK

Globally, the biotechnology industry is shifting its value creation emphasis from R&D to manufacture and marketing as more products emerge from the R&D pipeline and the industry as a whole matures. While we have looked at this broadly, we can explore some of the implications of this development on actual national industries. Due to such factors as industry concentration, resource access, sophistication of financial markets and quality of science, the industry has developed at a different pace in different nations. In evolutionary stages Australia trails behind the USA and Canada. The UK, through the Biotechnology and Biological Sciences Research Council (BBSRC), has built a strong biotechnology profile in a relatively short space of time, but based on a long history of fundamental biological research, epitomized by the discovery of the structure of DNA and pioneering work on determination of the three-dimensional structure of proteins in the 1950s.

Industry size – One distinguishing feature of the Australian industry is the small average size of its companies. While reports indicate that UK only has 400 biotechnology companies (Sainsbury 2002) to Australia's touted 300, its average company size in terms of both employees and turnover is significantly greater (Deloitte Touche Tohmatsu 2003). The UK companies are mature by Australian standards as most are publicly listed, have products to market and/or in the latter stage of clinical trials (Sainsbury 2002).

While there are (as of 2004) 105 listed Australian biotechnology companies, the industry is dominated by three players: CSL, Resmed and Cochlear, which account for 54 per cent of market capitalization (EG Capital 2004). Average capitalization of listed Australian biotechs is $5 million, making them on average quite small. This has a major impact on their ability to compete internationally (an issue which will be discussed in much more depth in Chapter 12) and to proceed to later stage development without further substantial investor, or indeed government, support. This is seen as one of the major limitations of the Australian industry, as well as a major reason for placing so much store in potential flagships such as Biota.

The difference in industry sector size and company life cycle in the UK and Australia is of significance, as Biotechnology Australia (2000) believes that Australia's small biotechnology sector has resulted in lost opportunities for commercialization through a lack of capital investment in Australia. In contrast, the UK's leading biotechnology sector is viewed as 'one of the best places in the world for bioscience' business opportunities (Sainsbury 2002, p. 20).

Strengths and Weaknesses of Each Industry

Both countries can boast supportive government policies, while neither has an identifiable focus on any sub-sector in the industry. Both countries have a strong publicly funded education system, strong educational and health infrastructures which have traditionally been publicly funded. Both have stable political environments with liberal ethical stances evident in their regulator environments.

To its detriment Australia has only three sizable biotech companies, the largest CSL with 8000 employees and a strong global market. Australia has been notable for its high productivity, particularly in R&D and has become known as an inexpensive place to do quality research. However, most pre-clinical and clinical trials are conducted offshore, as capacity and capability beyond applied research is currently limited. Aligned with this, in a chicken and egg dilemma, Australia also has an underdeveloped venture capital market, which will only emerge when their successful prodigies can offset the bad debts incurred from their forays into the IT boom, bubble and bust.

By contrast, the UK is a world leader in biotechnology with a mature investment community that complements their established science base and business infrastructure. However, in recent years an overall decrease in higher education spending has led universities to seek to commercialize outputs from their research projects.

The building of networks, and the comparative strengths and weaknesses between the UK biotechnology sector and that of Australia is summarized in Table 11.2 and 11.3. While the industry is global in operation and competition, there remain national differences based upon politics, age of the industry, skills and knowledge base, domestic demands and industrial orientation.

On the down side, Senker (2001) identified the lack of international competitiveness and consumer resistance to new products as a concern for the UK biotechnology industry as well. This is also an area which has caused some concern in Australia, to the extent that the major national biotechnology communication body, Biotechnology Australia, undertook

Table 11.2　Strengths of the UK and Australian biotechnology industries

Strengths	UK	Australia
Substantial public investment in research and development	✓	✓
Government support	✓	✓
Relatively low cost of research		✓
Developed research structure	✓	✓
Internationally well-regarded research capabilities	✓	✓
Global leader in biotech	✓	
Strong and experienced investment community	✓	
Megadiverse natural resources		✓
Easy access to European markets	✓	
Low taxation and inflation	✓	
Consumer friendly regulatory framework	✓	
Established science base and business infrastructure	✓	

Source:　Throssell, 2004, adapted from Finkel, 1999; Greek, 1999; Croft, 1999; Biotechnology Australia, 2000; Mercorp Consulting, 2000; Boguslavsky, 2000; NSW Government, 2001; Enright and Roberts, 2001; Sainsbury, 2002; Queensland Government and Ernst & Young, 2003.

Table 11.3   Weaknesses of UK and Australian biotechnology industries

Weaknesses	UK	Australia
Small biotechnology sector		✓
Lack of capital investment		✓
Innovation being taken offshore		✓
Lack of strong biotechnology manufacturing base to support	✓	
Transition phase from Research and Development to production		
Decrease in overall spending in higher education		✓
Lack of international competitiveness	✓	✓
Consumer resistance to new products	✓	

Source: Throssell, 2004, adapted from Finkel, 1999; Greek, 1999; Croft, 1999; Biotechnology Australia, 2000; Mercorp Consulting, 2000; Boguslavsky, 2000; NSW Government, 2001; Enright and Roberts, 2001; Sainsbury, 2002; Senker, 2001; Queensland Government and Ernst & Young, 2003.

a national study in 2002 to determine the main area of consumer reaction, and public concern. Unsurprisingly the big ethical issues such as cloning and xenotransplantation were major concerns to the public and were also the major issues raised in newspapers. In classic colonial style it was discovered by Biotechnology Australia that many of the editors of regional newspapers in Australia spent their early careers in Britain. As a result, to access information on various topics of public interest, they simply went back and read British newspapers (usually the tabloids) to gain their conservative, uninformed perspectives. The remedy was to educate the editors and provide access to real information on the major issues. Newspaper stories and editorials improved dramatically in the positive from that point.

The UK's 2002 Spending Review added another £1.4 billion to the science budget over three years to seize opportunities offered by bioscience and to maintain European leadership in biotechnology (Sainsbury 2002). Both the UK and Australia governments have demonstrated political support and a firm financial commitment to their respective biotechnology industries (see Table 11.4). However in Australia leadership has not come at the national level, with most support being devolved to the states.

The trends in Australia reflect changes in the biotechnology industry in the United States and worldwide, where the emergence of new classes of therapeutic and diagnostic molecules has created a bottleneck in the production of sufficient quantities of material for pre-clinical trials. Anecdotal

*Table 11.4   Alternative government and industry perspectives on
biotechnology growth paths through alliances*

Industry perspective	Government perspective
• Major academic institution at the heart of cluster	• Strong science base
• High tech industries	• Entrepreneurial culture
• Famous business school nearby	• Growing company base
• Access to capital	• Ability to attract key staff
	• Premises and infrastructure
	• Availability of finance
	• Business support services and large companies
	• Skilled workforce
	• Effective networking
	• Supportive policy environment

Source: Adapted from Persidis 1999; UK Department of Trade and Industry 1999;
Sainsbury 2002

evidence suggests that many Australian biotechnology companies are at a point of seeking to manufacture sufficient quantities of product to undertake preclinical trials. The capacity to undertake the required amount of manufacturing in Australia is limited (Hopper and Thorburn 2003), and unless domestic capacity to service the needs of these growing firms is established, the exodus of development opportunities to overseas contract research organizations (CROs) and pharmaceutical firms will continue.

Both countries will see a maturing of their companies. As medium-sized companies reach and succeed or fail with their Phase II trials, flagship companies will emerge which will influence the future direction of the industry in their respective countries. A clear example of the expectations of a company to succeed and lead the industry charge in Australia is Biota, whose major product Relenza® failed to achieve the market coverage and industry status expected after market launch. Within twelve months of market launch, licensee GSK had withdrawn most of its marketing resources, Relenza had been usurped by its main rival Tamiflu® and its share price had plummeted. The anticipated sun did not rise for Biota and the company that governments were seeking to identify an industry with, did not deliver all it promised. However like the biotech industry in Australia, Biota continues to trade, continues to research, develop products and engage in legal action.

Other likely influences on industry trends for the two countries include:

- Shifting government funding priorities will potentially have a major impact, particularly in terms of research programmes in universities and institutions, for example biosecurity is receiving significant funding boosts in both countries currently as a result of the so-called 'war on terror'.
- Shifts in proportions of GERD (gross expenditure on R&D) and BERD (business expenditure on R&D), as the industry matures, as discussed in Chapter 10.
- Increased industry concentration as companies mature and grow, the number of IPOs increases, M&As take effect and death rates amongst small firms impact.
- Non-biotech background entrepreneurs enter the market as the value bubble in the value chain moves away from research closer to the market (a warning here that the pendulum could well swing too far here as BERD takes over from GERD and market-focused entrepreneurs direct resources at the latter section of the value chain, thereby depriving the research phase of resources. This would potentially kill off the innovation and consequent product streams that the sustainability of the industry is set to rely upon).

CONCLUSION

NBFs play an important role in the biotechnology industry by their sheer weight of numbers, by their innovative endeavours and by their collaborations with larger firms and with institutions. At present the industry structures are creating advantage for small firms to emerge in many biotechnology industries around the world. However, the industry regimes they are born into are still quite rigid and the NBFs need to work within these environments if they are to succeed. Effective operation within the collaborative environments that exist, rather than working in isolation, is a recommended modus operandi. However, a constant awareness of the extent of change occurring within the industry in terms of the influence of institutional players, changing roles of government, growth of Big Biotech as well as the status of Big Pharma, must all be monitored. To do this, NBFs simply have to 'get strategic' as quickly as possible to fit with the industry regime and keep abreast of change.

REFERENCES

Ahuja, G. (2000), 'Collaboration networks, structural holes, and innovation: A longitudinal study', *Administrative Science Quarterly*, **45** (3), 425–57.

Barringer, B. and J. Harrison (2000), 'Walking a tightrope: creating value through interorganizational relationships', *Journal of Management*, **26** (3), 367–84.

Biotechnology Australia (2000), *Biotechnology Intellectual Property Manual*, Canberra: AGPS.

Bloch, C. (2004), 'Biotechnology in Denmark: a preliminary report', Danish Centre for Studies in Research and Research Policy working paper 2004/1, University of Aarhus.

Boguslavsky, J. (2000), 'Biotechnology sector thrives', *Research and Development*, **42** (9), 14–15.

Chandler, A. (1990), *Scale and Scope: The Dynamics of Industrial Capitalism*, Cambridge, Harvard University Press.

Coase, R. (1937), 'The nature of the firm', *Economica*, **4**, November, 386–405.

Croft, S. (1999), 'UK formulates long term research and development vision', *Research and Development*, **41** (10), 8–10.

Deloitte, Touche Tohmatsu (2003), *Borderless Biotechnology 2003*, New York: Deloitte, Touche Tohmatsu.

Doh, J. (2000), 'Entrepreneurial privatization strategies: Order of entry and local partner collaboration as sources of competitive advantage', *Academy of Management Review*, **25** (3), 551–71.

Doz, Y. (1996), 'The evolution of cooperation in strategic alliances: Initial conditions or learning processes?', *Strategic Management Journal*, **17**, 55–78.

EG Capital (2004), *Equity Research: Australian Biotechnology Expo 2004*, Sydney: EG Capital.

Enright, M. and B. Roberts (2001), 'Regional clustering in Australia', *Australian Journal of Management*, **26**, August, 65–84.

Ernst & Young (2003), *European Biotechnology Report*, London: Ernst & Young.

Etkowitz, H. and L. Leydesdorff (1999), 'The future location of research and technology transfer', *Journal of Technology Transfer*, **24** (2–3), 111–26.

Finkel, E. (1999), 'Budget backs report on boosting biotech', *Science*, **284** (5418), 1248–9.

Freeman, C. (1987), *Technology Policy and Economic Performance: Lessons from Japan*, London: Pinter.

Greek, D. (1999), 'Failing to make the best of our biotech', *Professional Engineering*, **12** (3), 14–15.

Gulati, R. (1998), 'Alliances and networks', *Strategic Management Journal*, **19** (4), 293–117.

Gulati, R., N. Nohria and A. Zaheer (2000), 'Strategic networks', *Strategic Management Journal*, **21** (3), 203–21.

Hitt, M., D. Ireland and R. Hoskisson (2001), *Strategic Management: Competitiveness and Globalization*, Cincinnati: South-Western College Publishers.

Hopper, K. and L. Thorburn (2003), 'Preclinical and scale-up manufacturing capacity in the Australian pharmaceutical industry – a comparative analysis', report for the Pharmaceutical Industries Action Agenda Implementation Team, Aoris Nova, Advance Consulting & Evaluation and BioAccent, Melbourne.

Ireland, D., M. Hitt and D. Vaidyanath (2002), 'Alliance management as a source of competitive advantage', *Journal of Management*, **28** (3), 413–26.

Jegathesan, J., A. Gunasekaran and S. Muthaly (1997), 'Technology development and transfer: experiences from Malaysia', *International Journal of Technology Management*, **13** (2), 196–215.

Landau, R. and N. Rosenberg (1986), *The Positive Sum Strategy: Harnessing Technology for Economic Growth*, Washington, DC: National Academy Press.

Lundvall, B. (1995), 'User-producer relationships, national systems of innovation and internationalisation', in B. Lundvall (ed.), *National Systems of Innovation and Interactive Learning*, London: Biddles Publishers.

Lundvall, B. (2002), *Innovation, Growth, and Social Cohesion: the Danish Model*, Cheltenham, UK and Northampton, MA, USA: Edward Elgar Publishing.

Mercorp Consulting (2000), 'Brisbane in formation technology and knowledge based industries clusters action plan', Brisbane: Mercorp.

Mitchell, G. (1999), 'Global technology policies for economic growth', *Technological Forecasting and Social Change*, **60** (3), 205–14.

Nonaka, I. and H. Takeuchi (1994), *The Knowledge-creating Company: How Japanese Companies Create the Dynamics of Innovation*, New York: Oxford University Press.

NSW Government (2001), *BioFirst: NSW Biotechnology Strategy, 2001*, Sydney: NSW Government.

Persidis, A. (1999), 'Biotechnology clusters', *Drug Discovery Today*, **4** (7), 297–8.

Polanyi, M. (1966), *The Tacit Dimension*, Garden City, NY: Doubleday.

Queensland Government and Ernst & Young (2003), *Queensland Biotechnology Report 2003*, Brisbane: Ernst & Young.

Sabel, C. (1995), 'Studied trust: building new forms of cooperation in a volatile economy', *Human Relations*, **46** (9), 1133–60.

Sainsbury, L. (2002), 'Strong Government support for UK biotech', *Biotechnology Investors' Forum Worldwide*, No. 2, pp. 20–22.

Senker, J. (2001), 'European biotechnology innovation system. Analysis of the bio-pharmaceuticals sector for EC TSER Project (SOE1-CT98-1117)', SPRU, University of Sussex.

Sivadas, E. and R. Dwyer (2000), 'An examination of organizational factors influencing new product success in internal and alliance-based processes', *Journal of Marketing*, **64** (1), 31–49.

Throssell, G (2004), 'How and why biotechnology clusters are formed in UK and Australia?', MPhil thesis, University of Queensland.

Tuma, E. (1987), 'Technology transfer and economic development: lessons of history', *Journal of Developing Areas*, **21** (4), 403–227.

UK Department of Trade and Industry (1999), 'Lord Sainsbury launches 10-point action plan for biotechnology clusters', Available: http://www.dti.gov.uk, Accessed: 15 September 2004.

Williamson, O.E. (1985), *The Economics of Capitalism: Firms, Markets, Relational Contracting*, New York: Free Press/Collier Macmillan.

12. Product development and innovation diffusion

INTRODUCTION

The generation of new and novel ideas (creativity) and the subsequent development and commercialization of these ideas (innovation) forms the basis of new product development from a generic framework. Creativity and innovation also form a continuum in new product development processes and therefore are important antecedents to the delivery of new products to market. This chapter first defines innovation and creativity and describes their relationship. It also presents a brief description of the management of ideas as they pass from the individual to the organization. The further development and implementation of these ideas is related to new product development and how innovation diffuses within and from outside the organization.

Innovation has been discussed extensively throughout this book. It is recognized that innovation can occur in any part of the organization and is not restricted to R&D functions. The term 'creativity' is sometimes interchanged with the word 'innovation'. A clear difference exists between the terms, however they are interrelated and linked (Majoro 1988; Barker 2002). Creativity is the ability or aptitude by which individuals or groups generate or conceive new ideas, or adapt existing concepts into new principles; while innovation is the practical application of such ideas towards meeting customer needs and organizational objectives in a more effective way, i.e. creating value from the customers' perspective.

Creativity is the precursor stage to the innovation process. The generation of ideas alone does not lead to a new innovation. Innovation is a dynamic process of developing or modifying ideas through effort (processing) into something that is novel and useful (Majoro 1988), as outlined in Figure 12.1.

Innovation theory owes much to the work of Schumpeter, who defined innovation as a new product, new production process or new organization form such as a merger or opening up of new markets (Schumpeter 1934). According to Schumpeter, innovation was an economic concept rather than a technological one. It does not matter how marvellous a technological

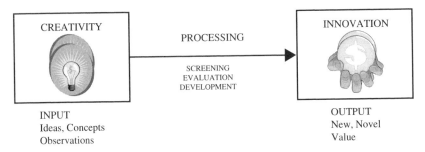

CREATIVITY

PROCESSING

SCREENING
EVALUATION
DEVELOPMENT

INNOVATION

INPUT
Ideas, Concepts
Observations

OUTPUT
New, Novel
Value

Figure 12.1   The relationship between creativity and innovation

invention may be, it is not an innovation unless it creates growth or profit in the market economy. Schumpeter also coined the term 'creative destruction' to describe the development of a new innovation. Schumpeter illustrated entrepreneurial innovation as a two-stage process: a discovery followed by diffusion or exploitation (Schumpeter 1934). He also noted that technological innovations appear in periodic cycles and are not evenly distributed over time or across industry (Schumpeter 1939). Drucker described innovation from the perspective of the organization. He defines innovation as the act that utilizes resources with the objective to create wealth (Drucker 1986). According to Drucker, innovation must be systematically driven.

INNOVATION MODELS

The process of innovation has always been a focus of researchers. Many models and hypotheses have been presented in the literature. Drucker (1986) describes the process of innovation as systematic and linear. This is the case for new product development, although concurrent or parallel processing also occurs (Smith and Reinertsen 1998). Organizational innovation has been described as a complex adaptive system (DISR 1999), similar to a living system adapting to changes in the environment to ensure survival.

Innovation has also been described as incremental, radical and transformational (Cooper 1994a). Tushman and Anderson (1986) debated whether innovation was continuous or discontinuous and whether it affects existing processes in an organization. An extension of this work described innovation as incremental improvements punctuated by discontinuous change (Tushman and O'Reilly 1997). Other authors have supported the view that innovation can be discontinuous (King and Anderson 1995; Holder and

Hamson 1995), that is, a break occurs in the continuity of a past innovation, which derives clearly from Schumpeter's early work on creative destruction.

A biotechnology example that demonstrates radical and incremental innovation is monoclonal antibody technology. When Kohler and Milstein discovered how to make monoclonal antibodies in 1975 they revolutionized the medical diagnostics and therapeutics industry sectors. For the first time, specific antibodies to various epitopes (targets) could be isolated and produced in significant quantities for applications in medicine. A number of incremental innovations built on the original approach established by Kohler and Milstein. These innovations included ease of producing the monoclonal antibodies, increasing the affinity of the antibodies to the specific target and reducing the development time. The monoclonal antibody procedure was still labour and resource-intensive, and still contained a large element of uncertainty in the product outcome. It wasn't until recently that another radical innovation allowed the production of monoclonal antibodies to move from the cellular to the molecular level. The ability to now produce monoclonal antibodies using molecular biology techniques has resulted in the establishment of significant monoclonal antibody libraries that can be screened for specificity across a number of targets. This range of innovations from incremental to radical is shown diagrammatically in Figure 12.2. A combination of the two rather than a reliance on radical innovation is a preferred path to organizational success.

Utterback and Abernathy (1975) explored the concept of dominant design and suggested that its occurrence may effect innovation in a firm or an industry. A dominant design is a new product synthesized from individual technological innovations introduced independently in prior product variants (Utterback 1994). The appearance of a dominant design provides an initial competitive advantage for the company introducing the product. It then has the effect of enforcing and encouraging standardization.

Disruptive innovation has been described by Christensen and Overdorf (2000) as a result of research studies conducted during the information technology boom of the 1990s. Disruptive innovation refers to a new product that does not initially address the needs of customers but becomes disruptive to the organization or industry. For example, the introduction of the polymerase chain reaction methodology by Roche was initially a disruptive innovation in biotechnology, in particular, related to molecular biology, genetics and medical diagnostics. PCR promised so much that it was perceived at the time of its introduction that it would displace a number of existing technologies, including enzyme-linked immunosorbent assay (ELISA), agglutination and other immunological tests. However, due to

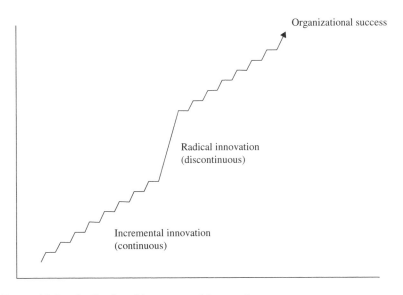

Figure 12.2   Radical and incremental innovation

the exorbitant and unrealistic licence fees imposed by Roche, the licence holder of the technology, this restricted the diffusion of this innovation throughout the biotechnology industry. In fact, it spawned a number of new innovations that emerged to bypass the PCR patent. Some of these technologies included ambient temperature amplification and rolling circle amplification. So PCR was not as disruptive as it could potentially have been because of limitations to its diffusion. A scientific success does not guarantee a market success.

Technology cycles first described by Schumpeter (1939) bring about discontinuities and unpredictability. A technological discontinuity, triggered by scientific or engineering advancement, ruptures existing incremental innovation patterns and spawns a period of ferment, where technological variants with different operating principles compete for market acceptance (Tushman and O'Reilly 1997). Competition also occurs between existing and new technology. This period of ferment causes uncertainty amongst customers and suppliers. The emergence of a dominant design results in a new standard that is accepted by the market. PCR has eventually become a dominant design where it is widely used in molecular biology, genetics, pharmocogenomics and medical diagnostics. Even though the PCR technology had a slow diffusion within the biotechnology industry, it became a dominant design through some successful licensing and distribution by

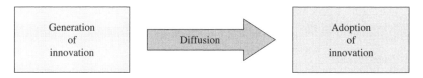

Figure 12.3 The stages of innovation

Roche, and because of the advantages it afforded to researchers, particularly in DNA research.

Furthermore, the expiration of the PCR patent will strengthen its dominance in the marketplace as the technology is freely diffused and adopted by biotechnology researchers and organizations. After the emergence of the dominant design, innovation shifts from major product innovation to process innovation and incremental, continuous innovation. Incremental innovation or continuous improvement (CI) is essentially a customer-driven process in most industries that facilitates the improvement of processes and products for the market. CI or building quality with products and processes provides the opportunity to develop a sustained competitive advantage. Finally a subsequent discontinuity triggers a new cycle of technological variation, selection and incremental change.

Innovation can be grouped into two broad stages: the generation of innovation and adoption of innovation (Gopalakrishnan and Damanpour 1997). However before the innovation can be adopted it needs to be diffused within the social system of the organization (Rogers 1995). Figure 12.3 shows a simple stage model for innovation.

Damanpour (1991) has a different view on the adoption of innovation, believing that innovation adoption encompasses each of generation, development and implementation of new ideas. The adoption stage is the acquisition and/or implementation of an innovation. The adoption of innovation generally contributes to the performance or effectiveness of the adopting organization.

CREATIVITY

Creativity is the ability or aptitude by which individuals or groups generate or conceive new ideas, or adapt existing concepts into new principles. For example, these ideas may be new solutions to a problem or a new method or product concept. By stimulating the creative process within individuals, new ideas and concepts can be generated that can lead to the achievement of goals.

The creative process was first described by Wallas in 1926. He proposed a systematic model that usually follows a sequence of phases outlined below:

1. Preparation
During the preparation phase relevant data and information is collected. The problem is defined and analysed and possible solutions are explored.

2. Incubation
When faced with a difficult problem sometimes it is best to sleep on it, which in turn frees the conscious mind from the problem. The subconscious mind continues to work on the problem through further analysis and synthesis of new combinations of ideas. The incubation phase provides the opportunity to reconsider the problem from a different perspective. It is believed that during the subconscious stage the right side of the brain starts to work on the problem in a holistic approach.

3. Illumination
The creative solution suddenly emerges into the conscious mind from the subconscious. This flash of inspiration is sometimes referred to as the 'Aha!' or 'eureka experience'. This dynamic moment usually occurs when you are not thinking about the problem, but are in a relaxed state.

4. Implementation/Verification
This stage is where the new idea, insight, intuition or solution is implemented and thoroughly tested. If the solution is not optimal further incubation may be required.

Creativity is often associated with the 'creative arts' such as music, painting and literature. The general perception of creative people is that they were born with unique creative skills. Michael Polanyi (1966) and Arthur Koestler (1964) began writing on the subject of creativity in the 1950s and 1960s further clarifying and describing tacit knowledge and creative thinking. Creativity and lateral thinking was later popularized by De Bono (1993) and Buzan (1995), reinforcing the idea that creativity was not innate but could be learned and deliberately applied to certain situations.

IDEA MANAGEMENT

It is well accepted that a continuous flow of ideas provides the antecedents for innovation within an organization (Majoro 1988), and in particular the development of new products (Kotler 1991). However, generating ideas

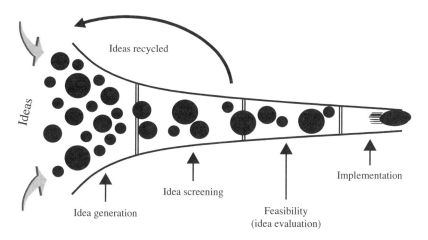

Figure 12.4   Idea funneling process

alone will not lead to innovation. Once ideas are generated they must be captured, screened, evaluated and finally implemented (Wheelwright and Clark 1995). Figure 12.4 outlines a typical idea funnelling process within an organization.

Ideas emanate both from internal sources (departmental/functional areas) and external sources such as competitors, customers, industry associations, research institutes, universities, the Internet, etc. Many organizations do not usually appreciate the wealth of ideas that surround them. Many new product/service ideas usually come from customers. By trapping the 'voice of the customer', organizations can tap into a readily available source of new product/service ideas.

The ability to encourage and exploit internal and external ideas without initial judgement is a key driver of innovation (Majoro, 1988). An organization must have a mechanism to capture (bank) these ideas before they dissipate and disappear.

In Chapter 5, we described the knowledge and idea management process in a unique way using the semi-permeable membrane. An idea management process encourages the active generation and collection of ideas. These ideas are then shared within the organization, usually by electronic means, and developed further using creativity tools and techniques. The ideas are then evaluated using an appropriate tool. This may involve conducting a technical or market feasibility study. The idea is finally harvested and implemented as a new product, new process or organizational innovation. Progress is monitored using measures to determine the effectiveness of the idea.

The idea management process begins with individual inspiration and the tacit knowledge of the individual. Tacit knowledge has been defined as non-codified, intangible know-how that is acquired through the informal adoption of learned behaviour and procedures (Howells 1996). Polanyi (1961) describes tacit knowing as involving two kinds of awareness: the focal and subsidiary. While individuals may be focused on a particular object or process, they also possess a subsidiary awareness that is subliminal and marginal (Howells 1996). A discovery that involves focused awareness is usually termed synchronicity since the individual is actively seeking an idea or a solution to a problem (Ayan 1997). Tacit knowing also involves subception, that is, learning without awareness and this is associated with serendipity. According to Jordan Ayan (1997) serendipity is defined as a random coincidence or accident that triggers an idea or concept when the individual is not actively seeking an idea i.e. without awareness of a problem or need.

Once the idea is generated, usually through a dynamic moment or illumination and is recorded, it becomes explicit or codified knowledge (Nonaka and Takeuchi 1994). The firm assimilates this knowledge through absorptive capacity (Cohen and Levinthal 1990).

Much of the essence of the innovation process is geared towards the economic benefit of product innovation. The internal process of bringing new ideas to fruition in product innovation is through new product development (NPD).

THE IMPORTANCE OF NEW PRODUCT DEVELOPMENT

New product development is critical to the success, survival, growth and renewal of organizations (Brown and Eisenhardt 1995). New products provide a steady stream of cash flow for business as a result of satisfying customer needs and wants. Organizations must also cater for the changing needs and wants of customers; otherwise competitors will try to fulfil these needs. New products provide a potential source of competitive advantage (Brown and Eisenhardt 1995) and allow organizations to diversify, adapt and even re-engineer to match involving markets and technical conditions.

The literature describes several examples of NPD processes. These processes range from relatively simple phase review systems to more complex stage-gate systems. These systems are described in this section together with a review of the evolution of NPD processes from simple first-generation to more complex third-generation processes.

In terms of first generation processes, the National Aeronautics and Space Administration (NASA) practised the concept of staged development in the 1960s with its phased project planning or what is often called 'phased review process'. The phased review process, regarded as a first-generation process, was intended to break up the development of any project into a series of phases that could be individually reviewed in sequence. Review points at the end of each phase required that a number of criteria be met before the project could progress to the next phase (Cooper 1994b). The aim of this process was centred on the physical design and development of the product and did not take into consideration the analysis of the external markets.

Another variation to NPD indicative of the old industrial linear R&D process was the departmental stage model. The NPD process was viewed in terms of departments that were responsible for the various tasks. For example, the R&D department provided the ideas and design, manufacturing produced the product, and then marketing became involved to launch the product to market. The process was designed in a 'pass the parcel' approach where a new project went from department to department, each department completing its own function (Hart and Baker 1994).

The stage-gate models originally developed by Cooper (1983) resembled the phased review process of the 1960s consisting of discrete stages preceded by review points or 'gates'. The main difference however, was that the stage-gate involved a cross-functional team of people (i.e. from R&D, marketing, manufacturing, quality, finance and administration), rather than a sequential passing the project along the different functions.

These second-generation systems involved activities from many different departments at each stage. Ownership of each stage transcends the whole organization. This approach helps reduce the influence and barriers created by functional empires, where one department may take control over the process.

Third-generation new product processes are further evolutions of second-generation stage-gate systems. These include further efficiencies in speed to market and allocation of development resources. Third-generation processes involve balancing speed to market with the need for thoroughness of information. According to Cooper (1994b), third-generation processes have four fundamental characteristics, which include:

1. *Fluidity* Stages are fluid (sometimes overlapping) and adaptable for greater speed.
2. *Fuzzy gates* Feature conditional 'Go'/'Kill' decisions based on the situation rather than absolute decisions.

3. *Focused* Resources are focused on the most promising projects in an entire portfolio rather than working on one project at a time.
4. *Flexibility* Since each product development project is unique the process is not a rigid stage-gate system.

'New generation' processes continue to emerge.

PORTFOLIO MANAGEMENT

An extension of third-generation processes is portfolio management. Portfolio management refers to the prioritization of a number of new product development projects through the allocation of appropriate resources – capital, labour and materials (Cooper et al. 1998).

Portfolio management is a tool used to outline the number of new product opportunities presented and the optimal investment mix between risk and return (Cooper et al. 1998).

Portfolio management is about making strategic choices for the organization – types of products, markets and technologies to target, and allocating and balancing scarce resources, such as R&D, marketing, operations and financial (Roussel et al. 1991).

STAGE-GATE MODEL

Stage-gate models with different variations have been presented by a number of publications (Booz et al. 1982; Cooper 1983, 1994b, 1990, 2001; Cooper and Kleinschmidt 1987; Crawford 1983; Kotler 1991). A typical stage-gate system used extensively is shown in Figure 12.5. The model incorporates the following components (Cooper 1990):

Idea generation
A new product idea is submitted to Gate 1 for review and initial screening.

Gate 1: initial screen
The initial decision to commit resources to the project. The project is assessed against key 'must meet' and 'should meet' criteria. The criteria include strategic alignment, project feasibility, magnitude of the opportunity, differential advantage, synergy with the firm's core business and resources, and market attractiveness (Cooper 1990). Financial criteria are not part of this early stage screening. Checklists with the appropriate criteria and weighting are used to rank the projects.

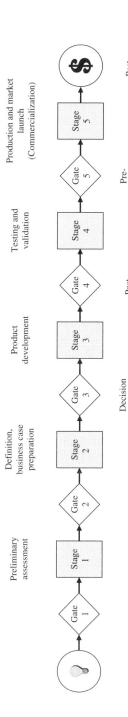

Source: Adapted from Cooper 1990

Figure 12.5   The Stage-gate process

Stage 1: preliminary assessment

Preliminary technical and marketing assessments are performed in a short period of time and with relatively low budget allocation. This is to obtain a quick assessment to determine the market potential and size, likely market acceptance, development and manufacturing feasibility and possible costing and timelines (Cooper 2001). This stage involves inexpensive activities such as a library search, contact with key customers and focus groups.

Gate 2: second screen

The project is re-evaluated in light of the new information obtained in Stage 1. The original 'must meet' and 'should meet' criteria are used together with any new criteria such as, customer reaction and feedback to the product. If the project receives a decision to 'Go' then it moves into the next stage.

Stage 2: product definition

The product at this stage must be clearly defined. Customer needs and wants are identified through market research studies, concept testing is performed with customers and a competitor analysis is undertaken (Cooper 1990). Detailed technical and operational appraisals are done to determine the technical and manufacturing capability of the organization. If manufacturing costs are shown to be too high then outsourcing may be investigated. A detailed financial analysis and legal/patent/copyright checks are undertaken.

Gate 3: decision on business case

This is the final decision point before the project enters into comparatively greater spending in the development stage. The project is again assessed using the 'must meet' and 'should meet' criteria with a greater reliance on the results of the financial analysis. Agreement must be reached by the gate-keepers on a number of key items, including target market definition, definition of the product concept, product positioning strategy and the product benefits that will be delivered to the customer (Cooper 2001). In addition, agreement must be made on the product definition, that is, the product features, attributes and specifications. It is important to lock the product definition at this point, as this will have an enormous bearing on the development plan and costs. Development plans and marketing plans with Gantt charts are reviewed and approved.

Stage 3: development

This stage involves the development of the product. Concurrently detailed product testing, marketing and operational plans are developed and the financial plans are updated.

Gate 4: post-development review

This gate involves a check on the progress of the project and the continued attractiveness of the product. The revised financial plan is reviewed to re-assess viability of the project. Testing and validation plans are reviewed and approved by the gatekeepers for immediate implementation in the next stage (Cooper 1990). Further review of the manufacturing and marketing plans is performed.

Stage 4: validation

The entire project is tested for its viability. This includes the product itself, the production process, customer acceptance and the financial require-ments of the project (Cooper 2001). This stage also includes the following activities (Cooper 1990):

- In-house testing to check product quality and performance
- Field/clinical trials to verify that the product works under user conditions, including gauging customer reactions
- Pilot production to test production process and determine more accurate costs
- Market test to determine the effectiveness of the product launch plans and to trial the selling approach.

Gate 5: pre-commercialization decision

This gate is the final decision point where the project can still be 'killed'. Financial projections play a key role in decision making. Final operations and marketing plans are reviewed and approved before implementation.

Stage 5: commercialization

The final stage involves the implementation of the market launch plan and operational plans.

Post-implementation review

Following commercialization, the new product, if successful, becomes a regular product line for the organization. Product and project performance are reviewed after the product is released to market. Real market revenue data, costs, expenditures and profits, including the timelines are compared against the original projections. A critical assessment of the project is also undertaken to determine the project's strengths and weaknesses, learnings from the project and improvements to be considered for future projects (Cooper 1990).

The stage gate model provides a disciplined approach to the new product development process. The process is visible and relatively simple. It provides a framework to facilitate NPD projects and give better definition of objectives, tasks and milestones. It is not the intention of the model that all projects pass through every stage. Some projects may be small or extensions of existing products that do not require rigorous staging and decision-making. Major projects involving investment will benefit from the stage gate model. A number of factors will determine whether a new project will progress through all the stages and gates of the model, including, the scope of the project, the investment required and the level of risk associated with the project (Cooper 1990).

An adaptation of the above model, previously described as a third-generation process, involves flexible, overlapping and fluid stages that are difficult to define. These emerging processes focus on speed and effective allocation of resources, allowing activities to progress to the next stage without total completion of the prior stage. Adaptability in the system overcomes the time-consuming and bureaucratic second-generation processes. Flexibility in activities rather than rigidity is promoted. Adaptability, fluidity, flexibility and focus are trends in management and strategy that have been described by other authors (Quinn 1992, Hamel and Prahalad 1994, Stevens et al. 1999). Cooper (1990) has incorporated these trends into the third-generation NPD models. Rothwell (1994) proposes a similar view in his fifth-generation innovation process model where he identifies integration, flexibility, networking and parallel (real time) information processing as the key characteristics of the process.

The end result of the stage gate model is commercialization. The commercialization process, while not involving the creative processes of the early innovation phases, is still an important aspect of innovation. Effective commercialization is enabled by the diffusion of the innovation, which requires its adoption in the market.

INNOVATION DIFFUSION AND ADOPTION

It is critical to the economic development of an industry that such developments spread quickly in order to maintain or enhance competitiveness. Despite limitations to the conceptualization of innovation diffusion we can utilize the underlying concepts to gauge idea spread in a region, market or industry. Once we know the idea spread, we can enquire into factors which will enhance or impede this spread. These are the barriers and enablers to diffusion.

The terms 'diffusion' and 'adoption' are used interchangeably in the literature (e.g. Baptista 2000), and have varying scope of meaning. Diffusion is a broad concept, being the 'way in which an innovation is spread through market and non-market channels' (OECD 1992). Kautz and Larsen (2000) see diffusion as the spread of knowledge, stating it 'largely is a communication process'.

The spread of an innovation throughout the market for which it was created is the way an innovation can be deemed successful. An innovation needs to have a performance enhancement or economic effect on those using it, compared to the way users had previously operated.

In understanding the characteristics, mechanisms and implications of innovation diffusion, reference must be made to the comprehensive work on the diffusion of innovations by Rogers (1995). Multidisciplinary studies in the areas of anthropology, sociology, education, industrial economics, medical sociology, and reviews of current research are generalized into the process of diffusion. Rogers (1995) proposed the four elements of diffusion as:

1. The innovation
2. Its communication from one individual (or entity) to another
3. Occurs in a social system
4. Transpires over time.

Variation in these elements can affect diffusion rates in many ways, especially in an organizational setting. For example, collective problem solving (in organizational adoption decisions), norms and the culture of the groups, roles and type of actors, presence of influencers (also known as change agents), and origin of the innovation and the influence of this (that is, the 'cosmopoliteness' of actors in accepting external innovations) (Rogers 1995).

Thus, the many elements involved in diffusion highlights the significance of communication and knowledge transfer to the innovation process and the diffusion of these innovations. However, understanding innovation adoption is vital to improving the innovation process, particularly as the adoption of an innovation is said to be a determinant of the innovation's success and hence its significance to the firm (OECD 1996). In other words, value is not found in its creation, but in its subsequent use.

Innovation adoption focuses on the decision to use an innovation and how managers come to that decision. For example, Klein and Speer Sorra (2001) define adoption as the manager's decision to use an innovation, which implies that managers have access to certain information before adoption.

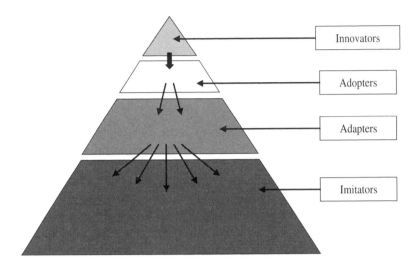

Figure 12.6   Hierarchy of adoption of innovations showing relative categories

The innovation-decision process has been defined as the process through which an individual (or other decision making unit) passes from:

1. First knowledge of an innovation
2. To forming an attitude toward the innovation
3. To a decision to adopt or reject
4. To implementation of the new idea, and
5. To confirmation of this decision. (Rogers 1995)

Whether the decision process is a logical, rational step-by-step process has been debated. However, research conducted to date has indicated that this pattern has consistently been shown to exist in varying degrees. In addition, outcomes of the decision process may be used to differentiate adopters, which can be categorized by comparing adoptions timings.

Five adopter categories that have emerged are: innovators, early adopters, early majority, late majority, and laggards. Figure 12.6 shows their relationship and the spread of innovations between adopters. Clearly there will be very few innovators, limited numbers of quick responding early adopters, then as the industry matures there will be an increasing majority of laggards.

These categories have similarities to marketing classifications for consumers in the adoption process. These labels are based on the time taken to adopt and thus each has their own characteristics of purchasing/adopting

Table 12.1   Adoption categories from institutional and marketing perspectives and identifiable groups in the biotechnology industry

Institutional	Marketing	Likely groups in the biotechnology industry
Innovators	Innovators	Star scientists
Adopters	Early adopters	NBFs and Big Biotech
Adapters	Early majority	Big Pharma
	Late majority	Out-of-industry competitors
Imitators	Laggards	Uninformed opportunists

behaviour (Rogers 1995). A link can be drawn between the marketing and institutional adoption categories as shown in Table 12.1.

Research on adopter categories based on time to adopt has led to further analysis of the adoption timing process. The measurement in these analyses is also less subjective. For example, the adoption decision has implications for 'innovation speed' – the elapsed time between an innovation's conception and commercialization (Kessler and Chakrabarti 1999). The factors that influence adoption decisions may accelerate or slow innovation speed, with implications for competitive advantage. These influential factors can be categorized as either those that inhibit (barriers) or spur (enablers) innovation adoption.

Technology diffusion in the biotechnology industry is a complex, iterative and often long-term process. It is about people, processes, cooperative and focused problem-solving. It involves changes at the strategic, management and operational levels and may involve breakthrough or radical change. However, more often it is about responding to diverse supply and demand problems by translating new and innovative ideas through discovery to commercialization. New and emerging technologies are key drivers in the biotechnology sector and will require new methodologies to deliver products and services to the marketplace. The vehicles to deliver these products and services to the market involve venture capital support, start-ups and spin-offs, collaborative and joint venture relationships, incubation, mentoring and other support systems.

From a wider economic development perspective, successful technology diffusion initiatives provide several major advantages and opportunities. First, they drive research and know-how to commercial outcomes. Second, by embedding new technologies into domestic firms this provides the best possible protection for locally developed commercial research. Finally, it is a primary vehicle for competing at a global, sectoral and firm level.

NETWORKS AND NETWORKING

The biotechnology industry has seen a number of acquisitions and mergers since 1999. One of the more significant mergers has been that of GlaxoSmithKline. Pfizer's acquisition of Pharmacia created a company with 12 blockbuster products and global sales of over US$40 billion. Although mergers and acquisitions create larger revenues and expand product portfolios, while establishing economies of scale, size alone will not overcome the productivity and profitability crisis in the pharmaceutical industry that has been demonstrated in recent data (Lloyd 2003). A larger workforce is not more efficient. Size only generates higher revenues, not necessarily higher returns, since revenues are directly proportional to sales, administration and general expenditure, while R&D investment is directly proportional to pipeline productivity.

Adopting a networked approach allows pharmaceutical companies to become more efficient by restructuring and outsourcing low value-add activities in the form of temporary and long-term strategic alliances, both at the domestic and international level. High value-add, critical activities such as maintaining the intellectual capital of the organization is kept in-house. Specialist biotechnology and smaller biopharmaceutical companies are more efficient and progressive than larger pharmaceutical

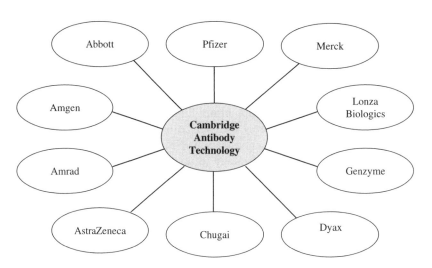

Source: Cambridge Antibody Technology 2005

Figure 12.7   Networks and partnerships of Cambridge Antibody Technology

companies. Outsourcing requirements and activities transfers a significant proportion of fixed costs to variable costs (Lloyd 2003). As a result of the changing nature of the industry, a networked pharmaceutical company can respond quickly and efficiently, without the constraint of having made in-house investments. The smaller, specialist biotechnology companies can reduce their risk or secure funding and resources by establishing network alliances with large pharmaceutical companies. Furthermore, these smaller biotech companies can increase their virtual size and remain sustainable through networks. Figure 12.7 shows some of the partnerships and networks established between Cambridge Antibody Technology (CAT) and pharmaceutical companies. The relationships include out-licensing, in-licensing and joint development of therapeutic antibodies.

REFERENCES

Ayan, J. (1997), *Aha!: 10 Ways to Free Your Creative Spirit and Find your Great Ideas*, New York: Three Rivers Press.

Baptista, R. (2000), 'Do innovations diffuse faster within geographical clusters?', *International Journal of Industrial Organisation*, No. 18, pp. 515–35.

Barker, A. (2002), *The Alchemy of Innovation: Perspectives from the Leading Edge*, Spiro Press, London.

Bessant, J., S. Caffyn and J. Gilbert (1996), 'Learning to manage innovation', *Technology Analysis and Strategic Management*, **18** (1), 59–70.

Booz, Allen and Hamilton (1982), *New Product Management for the 1980s*, New York: Booz, Allen and Hamilton.

Brown, S.L. and K.M. Eisenhardt (1995), 'Product development: past research, present findings and future directions', *Academy of Management Review*, **20** (2), 343–78.

Buzan, T. (1995), *Use Your Head*, London: BBC Books.

Cambridge Antibody Technology (2005), available at http://www.cambridgeantibody.com/html, Accessed on 2 April 2005.

Christensen, C.M. and M. Overdorf (2000), 'Meeting the challenge of disruptive change', *Harvard Business Review*, **78** (2), March, 66.

Cohen, W. and D. Levinthal (1990), 'Absorptive capacity: A new perspective on learning and innovation', *Administrative Science Quarterly*, **35**, 128–52.

Cooper, R.G. (1983), 'A process model for industrial new product development', *IEEE Transactions on Engineering Management*, EM-30, no.6, 2–11.

Cooper, R.G. (1990), 'Stage-gate systems: a new tool for managing new products', *Business Horizons*, **33** (3), 44–54.

Cooper, R.G. (1994a), 'New products: the factors that drive success', *International Marketing Review*, **11** (1), 60–76.

Cooper, R.G. (1994b), 'Third-generation new product processes', *Journal of Product Innovation Management*, **11**, 3–14.

Cooper, R.G. (2001), *Winning at New Products: Accelerating the Process from Idea to Launch*, 3rd edn, Cambridge, MA: Perseus Publishing.

Cooper, R.G. and E.J. Kleinschmidt (1987), 'New products: what separates winners from losers?', *Journal of Product Innovation Management*, **4**, 169–84.

Cooper, R.G., S.J. Edgett and E.J. Kleinschmidt (1998), *Portfolio Management for New Products*, New York: Addison-Wesley.

Crawford, C.M. (1983), *New Products Management*, Homewood, IL: Irwin.

Damanpour, F. (1991), 'Organizational innovation: a meta-analysis of effects of determinants and moderators', *Academy of Management Journal*, **34** (3), 555–90.

De Bono, E. (1993), *Lateral Thinking*, London: Penguin Books.

Department of Industry, Science and Resources (1999), 'Shaping Australia's future: innovation – framework paper', Department of Industry, Sciences and Resources Report.

Drucker, P.F. (1986), *Innovation and Entrepreneurship*, London: Pan Books.

Gopalakrishnan, S. and F. Damanpour (1997), 'A review of innovation research in economics, sociology and technology management', *Omega, International Journal of Management Science*, **25** (1), 15–28.

Hamel, G. and C.K. Prahalad (1994), *Competing for the Future*, Boston, MA: Harvard Business School Press.

Hart, S.J. and M.J. Baker (1994), 'The multiple convergent processing model of new product development', *International Marketing Review*, **11**, 77–92.

Holder, R.J. and N. Hamson (1995), 'Requisite for future success – discontinuous improvement', *Journal for Quality and Participation*, September, 40–45.

Howells, J. (1996), 'Tacit knowledge, innovation and technology transfer', *Technology Analysis and Strategic Management*, **8** (2), 91–106.

Kahn, K.B. (2001), *Product Planning Essentials*, Thousand Oaks, CA: Sage Publications.

Kautz, K. and E. Larsen (2000), 'Diffusion theory and practice: disseminating quality management and software process improvement innovations', *Information Technology and People*, **13** (1), 11–20.

Kessler and Chakrabarti (1999), Innovation speed: a conceptual model of context, antecedents, and outcomes', *Academy of Management Review*, **21** (4), October, 1143–92.

King, N. and N. Anderson (1995), *Innovation and Change in Organizations*, New York: Routledge.

Klein, K., A. Conn and J. Speer Sorra (2001), 'Implementing computerized technology: an organizational analysis', *Journal of Applied Psychology*, **86** (5), 811–29.

Koestler, A. (1964), *The Act of Creation*, New York: Macmillan.

Kotler, P. (1991), *Marketing Management: Analysis, Planning, Implementation and Control*, 7th edn, Englewood Cliffs, NJ: Prentice-Hall.

Lloyd, W. (2003), 'Editorial: A new era of networking', *Journal of Commercial Biotechnology*, **9** (2), 97–8.

Majoro, S. (1988), *The Creative Gap: Managing Ideas for Profit*, London: Longman.

Nonaka, I. and H. Takeuchi (1994), *The Knowledge-creating Company: How Japanese Companies Create the Dynamics of Innovation*, New York: Oxford University Press.

OECD (1992), *OECD Proposed Guidelines for Collecting and Interpreting Technological Innovation – Oslo Manual*, Paris: OECD.

OECD (1996), *OECD Proposed Guidelines for Collecting and Interpreting Technological Innovation – Oslo Manual*, Paris: OECD.

Polanyi, M. (1961), 'Knowing and being', *Mind*, **70** (280), 458–70.

Polyanyi, M. (1966), *The Tacit Dimension*, London: Routledge and Kegan Paul.

Quinn, J. (1992), *Intelligent Enterprise*, New York: Free Press.

Rogers, E.M. (1995), *Diffusion of Innovations*, 4th edn, New York: Free Press.

Rothwell, R. (1994), 'Towards the fifth generation innovation process', *International Marketing Review*, **11** (1), 7–31.

Roussel, P.A., K.N. Saad and T.J. Erickson (1991), *Third Generation R&D: Managing the Link to Corporate Strategy*, Boston, MA: Harvard Business School Press.

Schumpeter, J. (1934), *The Theory of Economic Development*, Cambridge, MA: Harvard University Press.

Schumpeter, J. (1939), *Business Cycles: A Theoretical, Historical and Statistical Analysis of the Capitalist Process*, New York: McGraw-Hill.

Smith, P.G. and D.G. Reinertsen (1998), *Developing Products in Half the Time: New Rules, New Tools*, New York: John Wiley.

Stevens, G., J. Burley and R. Divine (1999), 'Creativity + business discipline = higher profits faster from new product development', *Journal of Product Innovation*, **16**, 455–68.

Tushman, M. and P. Anderson (1986), 'Technological discontinuities and organizational environments', *Administrative Science Quarterly*, **31** (3), 439–65.

Tushman, M.L. and C.A. O'Reilly (1997), *Winning Through Innovation: A Practical Guide to Leading Organizational Change and Renewal*, Boston, MA: Harvard Business School Press.

Utterback, J.M. (1994), *Mastering the Dynamics of Innovation*, Boston, MA: Harvard Business School Press.

Utterback, J.M. and W.J. Abernathy (1975), 'A dynamic model of product and process innovation', *Omega*, **3** (6), 639–56.

Wallas, G. (1926), *The Art of Thought*, London: Jonathan Cape.

Wheelwright, S.C. and K.B. Clark (1995), *Leading Product Development: The Senior Manager's Guide to Creating and Shaping the Enterprise*, New York: Free Press.

13. Biotechnology industry growth models: an international perspective

INTRODUCTION

This chapter explores the challenge of taking the biotechnology industry forward, using a range of countries as examples, including Singapore, Australia, Europe and the USA. Development of the biotechnology industry is discussed, in conjunction with a consideration of governance, growth through biotechnology clusters, outsourcing for growth, and market share versus market size.

Models of industry formation are considered from the perspective of other industries and different countries. Facilitating industry growth, which is directly or indirectly the desire of most biotechnology related organizations, and removing current and potential barriers to growth, while maintaining an eye on public benefit, is problematic for both government and industry. Leadership is a critical question – should it be national governments, major research funding bodies, public research organizations, major companies, groupings of smaller companies, or a mixture of each? This is as yet unresolved in many countries. However, in Singapore, the path seemed clearer, given its small population and geographic space, and its single-tiered centralized government (although this was also thought to be the case for IT).

Although numerous models exist of how to create or build a biotechnology industry worldwide, a simple transfer of strategy from one geographical area or industry sector to another is unlikely to be successful. Each individual biotechnology company has a role to play as industry growth is the common desire of all biotechnology organizations. How it is achieved is an unknown. This chapter presents a range of ideas and alternative paths to growth and influences on growth for the reader to consider.

THE GLOBAL BIOTECHNOLOGY INDUSTRY

Biotechnology is an enabling technology and therefore has been embraced by governments to drive economic growth and improve quality of human life. The global biotechnology sector is currently undergoing change. The recent market downturn in the global economy is impacting on the current biotechnology landscape. The biotechnology sector is moving towards an alliance network of specialty companies (Ernst & Young 2003a). A company's activities do not have to span the full breadth of the biotechnology value chain. Instead, biotechnology companies focus on their core competencies or expertise. Companies however have to focus on areas that provide the greatest potential return or profitability through bringing products to market.

The global biotechnology industry consists of more than 4700 companies, 611 of these are publicly-traded companies, which achieved a net loss of more than US$4.5 billion in 2003 (Ernst & Young 2004). These public biotechnology companies employed a total of 195 820 employees and achieved combined revenues of US$46.5 billion in 2003. R&D investment was comparatively high with US$18.6 billion invested. Table 13.1 shows the breakdown of the global biotechnology. The biotechnology industry is analogous to the information technology industry – populated by a large number of entrepreneurial high technology firms (Deeds et al. 1999). As a consequence the industry is characterized by continually changing technologies and intense competition.

Companies worldwide are looking at consolidation and contraction, through merger or restructuring. Many companies that are running out of capital reserves are struggling to remain viable. Table 13.2 shows a slight reduction in the number of publicly-listed companies worldwide despite the 17 per cent increase in revenue between 2002 and 2003. However, the table also shows that net losses have decreased by 65 per cent from US$12.8 billion in 2002 to US$4.5 billion in 2003, indicating a positive turnaround in profitability.

Table 13.1   Global biotechnology at a glance in 2003

	Global	USA	Europe	Canada	Asia/Pacific
Public company data					
Revenues (US$m)	46 553	35 854	7 465	1 729	1 506
R&D expenses (US$m)	18 636	13 567	4 233	620	217
Net loss (US$m)	4 548	3 244	548	586	170
Number of employees	195 820	146 100	32 470	7 440	9 810
Number of companies					
Public companies	611	314	96	81	120
Private companies	3 860	1 159	1 765	389	547
Public and private companies	4 471	1 473	1 861	470	667

Source: Ernst & Young 2004

Table 13.2   Global biotechnology data comparison

	2003	2002	change (%)
Public company data			
Revenues (US$ million)	46 553	39 783	17
R&D expenses (US$ million)	18 636	22 074	−16
Net loss (US$ million)	4 548	12 835	−65
Number of employees	195 820	179 050	9
Number of companies			
Public companies	611	619	−1
Private companies	3 860	3 749	3
Public and private companies	4 471	4 368	2

Source: Ernst & Young 2004

The twelve largest biotechnology countries (based on the number of companies) are shown in Figure 13.1. The USA has the largest number of biotechnology companies, including the greatest revenues generated by public companies. For a comparatively small country, Canada has the next largest number of biotechnology companies after the USA.

USA

The global biotechnology industry is dominated by the USA, contributing to more than 70 per cent of revenues and more than 70 per cent of R&D spending (Ernst & Young 2003a). The industry in the USA is driven by a number of factors (Biotechnology Industries Taskforce 2000), including:

- A large research and knowledge base that is world class
- A strong entrepreneurial culture
- Accessibility to greater amounts of venture capital
- The presence of existing large multinational companies in the pharmaceutical, medical diagnostics and agricultural sectors
- Regulatory environment that is more demanding than other countries.

Revenues increased from US$33.6 billion in 2002 to US$39.2 billion in 2003 (revenue growth of 16.7 per cent) clear indication that the industry is going through an expansive cycle (Table 13.3). Employment figures between 2002 and 2003 remained stable, despite the growth in revenue.

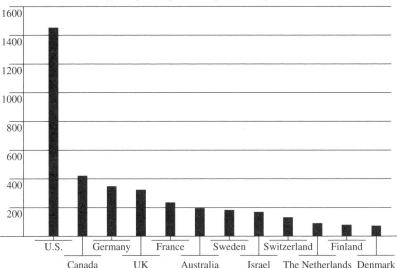

Number of biotechnology companies (public and private companies)

Source: Ernst & Young 2003a

Figure 13.1 Top 12 biotechnology countries

The majority of US biotechnology companies focus on the medical and health sector. However, it is predicted that the agricultural biotechnology sector will grow at 55 per cent per annum (Biotechnology Industries Taskforce 2000). As is the case in the global biotechnology industry, the industry in the USA is also undergoing consolidation, with a number of mergers and acquisitions occurring, as indicated by the fall in the number of public companies and their respective market capitalization (Table 13.4). Despite consolidation, the long-term potential for the industry remains, with biopharmaceuticals expected to grow at 12 per cent annually between 2003 and 2005 driven by the anticipated launch of 20 or more drugs per year through to 2004 (Rhodes 2002).

Following is a list of some industry facts and figures that describe activity in the US biotech industry (Biotechnology Industry Organization 2004):

- There are 1473 biotechnology companies in the United States, of which 314 are publicly held.
- Market capitalization, the total value of publicly traded biotech companies (USA) at market prices, was \$311 billion as of mid-March 2004.

Table 13.3   US biotechnology total company statistics (1998–2003)

	Total companies					
	1998	1999	2000	2001	2002	2003
Industry						
Number of companies	1311	1274	1379	1457	1466	1473
Employees	155 000	162 000	174 000	191 000	194 600	196 000
Sales (US$ billion)	14.5	16.1	18.1	21.4	24.3	28.4
Revenues (US$ billion)	20.2	22.3	25.0	28.5	33.6	39.2
R&D (US$ billion)	10.6	10.7	13.8	15.7	20.5	17.9
Net loss (US$ billion)	4.4	5.6	5.8	6.9	11.6	5.4

Source: Ernst & Young 2004

Note: Market capitalization not available; Revenues includes product sales and service revenue.

Table 13.4   US biotechnology public company statistics (1998–2002)

	Public companies				
	1998	1999	2000	2001	2002
Industry					
Number of companies	316	301	339	342	318
Employees	106 000	114 000	128 000	141 000	142 900
Market capitalization (US$ billion, 31 Dec.)	137.9	353.5	330.8	255	189.5
Financials					
Sales (US$ billion)	12.0	13.6	15.9	19.1	21.9
Revenues (US$ billion)	16.6	18.8	22.0	25.32	30.3
R&D (US$ billion)	6.7	6.9	9.9	11.53	16.3
Net loss (US$ billion)	1.9	3.2	3.9	4.8	9.4

Source: Ernst & Young 2003a

- The biotechnology industry has mushroomed since 1992, with US revenues increasing from $8 billion in 1992 to $39.2 billion in 2003.
- The US biotechnology industry employed 196 000 people as of 31 December 2003.
- There are more than 370 biotech drug products and vaccines currently in clinical trials targeting more than 200 diseases, including

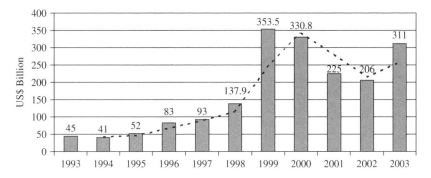

Source: Biotechnology Industry Organization 2004

Figure 13.2 Market capitalization in the US biotech industry

various cancers, Alzheimer's disease, heart disease, diabetes, multiple sclerosis, AIDS and arthritis.

- Biotechnology is responsible for hundreds of medical diagnostic tests such as those for diabetes, HIV (to keep the blood supply safe) and other diagnostic tests that detect conditions early enough to be successfully treated.
- The top eight biotech companies spent an average of $104 000 per employee on R&D in 2003.

Market capitalization and sales and revenues for the biotech industry in the USA give us some indications about the economic activity (Figure 13.2).

It can be observed in Figure 13.2 that the biotech industry had a slowdown from 2001–03, mostly due to economic uncertainty. But the figures for 2004 show a strong recovery and the future outlook is promising.

Much of biotech's outstanding performance in 2003 can be attributed to the continued shift from Internet and telecommunication stocks. Since the 'dot-com' crash, investors have been diversifying their portfolios away from technology heavy investments and increasing funding to life sciences concerns. The trend also reflects the traditional behaviour of investors looking to healthcare as a recession-proof haven in an uncertain market (PricewaterhouseCoopers 2004).

Other reasons explaining the biotech comeback include the emergence of new cancer therapies, a more efficient Food and Drug Administration, and a surge of merger and acquisition activity. For example, one of the major opportunities in drug development today is cancer, where currently marketed drugs have poor efficacy and high toxicity. Even a slight benefit or fewer side effects has the potential to generate a blockbuster drug. It is not surprising then that the sharp rise in biotech stocks began in May 2004 when encouraging news about several cancer therapies surfaced. This included Millennium Pharmaceuticals when it received FDA clearance to market Velcade, a treatment for multiple myeloma, after only four months of review. Millennium is now hoping to expand on the utility of Velcade with data in other indications. The Velcade approval was followed closely by the release of promising data on two colon cancer candidates: Genentech's Avastin™ and ImClone Systems' Erbitus™.

Canada

It is worth highlighting the biotechnology industry in Canada because after the USA it is the next largest in terms of number of companies. Canada has over 400 biotechnology companies and the industry is worth more than US$1.4 billion. Canada's industry is mainly focused on healthcare, however, it also has a strong agricultural biotechnology sector. Canada is well positioned for strong growth (Rhodes 2002) through:

- Resurgence in venture capital and government funding
- Strong research-based universities
- Continued Canadian success stories.

One of the biggest challenges faced by Canadian biotechnology companies has been the availability of venture capital opportunities, especially after the downturn in life-science investment that spilled over from the USA. Many of the Canadian companies are younger and smaller than US companies, therefore are more dependent on venture capital (Rhodes 2002). The government has had a positive influence through the facilitation of start-up venture funds, incubators and research centres. Table 13.5 outlines a summary of the Canadian biotechnology industry statistics.

Europe

The European biotechnology industry experienced rapid growth between 2000 and 2001 (Table 13.6). However, since 2002 growth has stalled due to the global economic downturn (Ernst & Young 2003b). Europe has 1878 biotechnology companies employing 82 124 people and generating €12.8 billion in revenue. The European industry as a whole spends €7.6 billion on R&D and recorded net losses of €4.0 billion in 2002 (Table 13.6). European biotechnology companies contribute approximately 20 per cent of global revenues and 25 per cent of R&D spending (Ernst & Young 2003b).

Table 13.5   Canadian biotechnology company statistics (2001–03)

	2001	2002	2003
Number of public companies	85	85	81
Number of public and private companies	416	417	470
Employees	7005	7785	7440
Revenues (US$ million)	1021	1466	1729
R&D expenditure (US$ million)	474	555	620
Net loss (US$ million)	507	263	586

Source: Ernst & Young 2003a, 2004

Table 13.6   European biotechnology company statistics (1998–2002)

	1998	1999	2000	2001	2002
Number of companies	1178	1352	1570	1879	1878
Germany/UK	222/268	279/275	332/281	365/307	360/331
Employees	45000	57589	61104	87182	82124
Revenues (€ billion)	3.7	6.285	8.679	13.7	12.9
R&D (€ billion)	2.3	3.364	4.977	7.5	7.7
Net loss (€ billion)	2.107	1.108	1.570	1.5	4.0

Source: Ernst & Young 2003b

Germany has the largest number of biotechnology companies with approximately 360 followed by the United Kingdom with 331 companies (Figure 13.3). However the UK, followed by Switzerland dominate European revenue contributions with approximately €3.0 billion and €1.8 billion respectively.

The outlook for the European biotechnology industry looks positive with approximately 456 products in the pipeline that are between pre-clinical and phase III trials. Leading up to 2001, a number of mergers and acquisitions occurred between European and US companies, resulting in contraction and consolidation of the industry.

Asia

A number of Asian countries are focusing their efforts on building biotechnology industries. The Japanese government anticipates the workforce

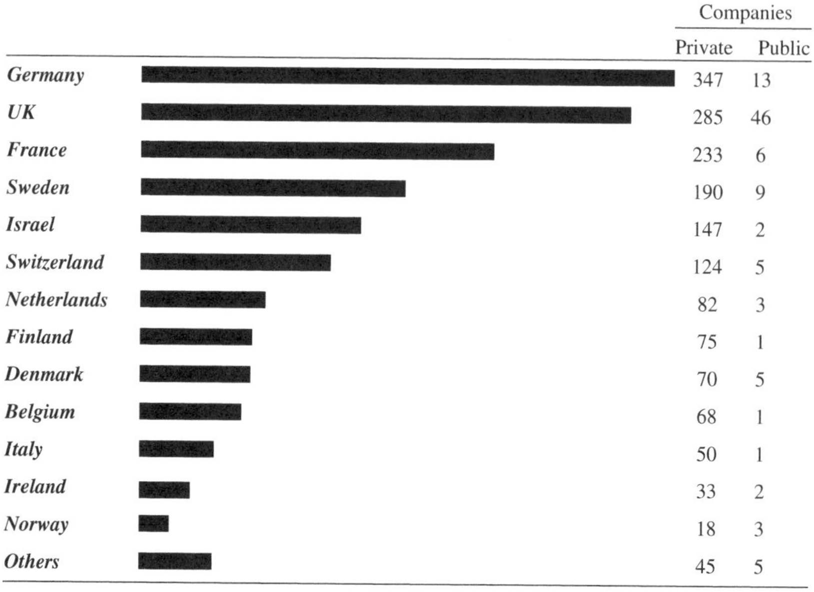

		Companies	
		Private	Public
Germany		347	13
UK		285	46
France		233	6
Sweden		190	9
Israel		147	2
Switzerland		124	5
Netherlands		82	3
Finland		75	1
Denmark		70	5
Belgium		68	1
Italy		50	1
Ireland		33	2
Norway		18	3
Others		45	5

Notes: Others refers to European countries surveyed by Ernst & Young whose numbers are too small to register by individual country.

Source: Ernst & Young 2003b

Figure 13.3   Biotechnology companies by European country 2002

involved in biotechnology related activities will surge to 1 million by 2010, over the current 70 000 employees (Ernst & Young 2003a). The government has a vision to create a $US208 billion biotechnology market by 2010 (Biotechnology Industries Taskforce 2000). The market size in 2001 was approximately US$10.5 billion (Rhodes 2002). The trigger behind the government's strategy is the perception that European and US firms are outperforming their Japanese rivals in the commercialization of biotechnology research (Ernst & Young 2003a).

The Singapore government has been aggressive in funding biotechnology initiatives and recruiting top biotechnology scientists (Biotechnology Industries Taskforce 2000). Singapore's revenues from biotechnology commercialization are expected to reach US$7 billion by 2007 (Ernst & Young 2003a).

The Chinese government has spent US$180 million between 1996 and 2002 to build its biotechnology industry. It is planning to spend more than three times that amount in the next three years (Ernst & Young 2003a).

India also has an emerging biotechnology industry and in the years to 2010 is expected to generate US$5 billion in revenues and 1 million new jobs (Ernst & Young 2003a). The government is promoting the biotechnology industry by providing infrastructure such as technology parks and by establishing technology transfer offices to facilitate technology licensing (Biotechnology Industries Taskforce 2000).

The Australian Biotechnology Industry

Background and industry highlights
Australia has a strong international reputation for scientific research and discoveries, however, it has lacked the ability to successfully commercialize many of these discoveries. One of the key deficiencies in the Australian biotechnology industry has been a lack of true risk capital. However, this is improving. Other factors that inhibit successful commercialization of scientific research include a lack of commercialization expertise, a general tendency of Australians to be risk-averse and a non-conducive tax environment.

The Australian biotechnology industry is dominated by small to medium enterprises (SMEs), with a small number of larger organizations such as CSL Ltd, ResMed Inc and Cochlear Ltd. (Ernst & Young 2001). The industry also includes subsidiaries of most international pharmaceutical and agribusiness companies, with many contributing to R&D and manufacturing investment in Australia. Alliances and joint ventures provide Australian biotechnology companies with a major route to international markets.

The Australian biotechnology industry is experiencing continued growth and expansion into international markets. Despite the industry being in a growth phase it remains small in global terms, with revenue generated by core biotechnology companies estimated at almost US$1 billion (Ernst & Young 2001). It is estimated that there are over 650 Australian biotechnology companies of which 236 are regarded as the core (Ernst & Young 2001, 2003a). Table 13.7 outlines the key statistics of the Australian biotechnology industry.

In 2001 a number of significant events occurred in the Australian biotechnology industry. Approximately 60 biotechnology companies listed on the Australian Stock Exchange. Of the 60 biotechnology companies, 35 were core and 25 were biotechnology-related. The major sector of activity is health (47 per cent) followed by agriculture, equipment/services, genomics/proteomics and bioinformatics (Figure 13.4). The health sector can be further broken down to pharmaceutical (63 per cent), diagnostics (15 per cent) or both (12 per cent) (Ernst & Young 2003a).

Table 13.7   Australian biotechnology at a glance

	2003	2002	change (%)
Public company data			
Revenues (US$ million)	980.6	920.5	7
R&D expenses (US$ million)	141.2	125.8	12
Net loss (US$ million)	110.9	71.0	56
Number of employees	6393	6464	−1
Market capitalization (US$ million)	4995.3	4134.8	21
Number of companies			
Public companies	58	50	16
Private companies	168	164	2
Public and private companies	226	214	6

Source: Ernst & Young 2004

The industry in Australia is supported by a number of public investment programmes focusing on R&D. Support from the federal and state government provides initiatives encouraging infrastructure development and facilitating collaborative alliances.

Research and development
Publicly listed Australian biotechnology companies spent about US$126 million on R&D in 2002 (Table 13.7) compared to US$16.3 billion by US public companies. This makes R&D expenditure comparatively low in global terms. Australian companies are limited by the available cash resources required to maintain research and carry out expensive clinical trials. Funding for R&D is an ongoing challenge for small to medium companies in Australia. This increases the company's risk profile and forces the company to focus on a limited portfolio of products to minimize cash burn rates.

The Australian government supports the industry by providing a significant amount of funding for R&D initiatives in Australia. In 2001 9.3 per cent of Australia's public sector R&D funding was committed to biotechnology (Ernst & Young 2001). Table 13.8 shows the estimated annual R&D commitment by the Australian government in biotechnology.

Products and pipelines of products
It is estimated that 51 per cent of Australian biotechnology products in the product pipeline are in the development phase, while 21 per cent are undergoing clinical trials or field tests (Ernst & Young 2001). The remaining

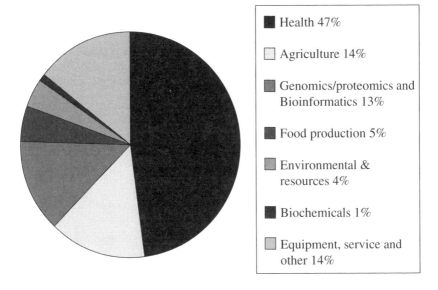

■	Health 47%
□	Agriculture 14%
▨	Genomics/proteomics and Bioinformatics 13%
■	Food production 5%
▨	Environmental & resources 4%
■	Biochemicals 1%
▨	Equipment, service and other 14%

Source: Ernst & Young 2001

Figure 13.4   Australian biotechnology industry sectors

28 per cent of products are on the market at various stages of their product lifecycles. The product pipeline is dominated by the four large companies (CSL Ltd, ResMed Inc, Cochlear Ltd, FH Faulding & Co), which account for 60 per cent of the products.

As of 2005, the product development pipeline is dominated by the human health sector with over 270 products in the pipeline. This is followed by the equipment and services sector and the agriculture sector, with approximately 150 products and 90 products respectively.

Competitive position summary
The competitive position of the Australian biotechnology industry is summarized in Table 13.9.

GLOBALIZATION

Globalization is defined as the expansion of global linkages, the organization of social life on a global scale, and the growth of global consciousness, all contributing to a single global economy.

Table 13.8   Estimated annual R&D commitment by the Australian government

	Biotechnology (AUD$m)	Total (AUD$m)	Share of total funding (%)
CSIRO	47	600	7.8
National Health and Medical Research Council	54	321	16.8
Australian Research Council	25	468	5.3
Co-operative Research Centres	28	134	20.9
R&D Start	17	155	11
Rural R&D corporations	14	170	8.2
Pharmaceutical Industry Investment Program	3	10	30
Universities	90	1320	6.8
Other	29	122	23.8
Total	307	3300	9.3

Note: 'Other' may include specific medical and other research institutes and individual research programmes.

Source: Ernst & Young 2001

Globalization involves three main features that seem to be the main engine driving world economic integration (Mrak 2000):

● Internationalization of production accompanied by changes in the production's structure.
● Expansion of international trade in trade and services.
● Widening and deepening of international capital flows.

Furthermore, the influence of the rapid development of science and technology and the ability to transfer it is driving a new global economic form. Many countries and regions are now beginning to recognize the importance of intellectual property (IP) and its protection as a building block in the pharmaceutical and biotechnology industries. Many of these nations will themselves be major contributors to new ideas and innovation. In 2003 India, China and the Asia-Pacific began to make meaningful intellectual contributions to the global biotechnology community.

The global biotechnology industry shows enormous potential for the future despite the recent economic downturn in the market. It promises to

Table 13.9   Competitive position of the Australian biotechnology industry

Strengths and opportunities
- Australia has a globally competitive industry
- Strong R&D base with substantial public investment
- A relatively low cost R&D structure
- Strong international reputation for quality of science and source of intellectual capital
- A well-educated/skilled and readily available labour pool
- Government (both federal and state) provision of significant funding and support of industry and infrastructure,
- Industry incentives in biotechnology R&D (e.g. Commercial Ready Grants and R&D tax concessions)
- Emphasis on high-value Australian-based R&D platform technologies e.g. stem cells, reproductive technology, proteomics
- Biodiverse ecosystems
- Initial successes in the biotechnology sector to build from
- Internationally regarded research capabilities
- Existing strengths in medical, agricultural and pharmaceutical R&D with commercial applications
- Strong regulatory framework with continuing harmonization with US and European standards

Weaknesses and threats
- Distance from major world markets, such as the USA and Europe
- Limited size of local markets due to comparatively low population
- Generally low R&D expenditure by the corporate sector and continued low investment by Australian business in biotechnology R&D
- Lack of industry-credible CEOs/CFOs with international commercial experience in the biotechnology industry
- Limited but improving entrepreneurial culture in academic and graduate communities
- Overall lack of scale in the biotechnology industry
- Small industry limits the mobility of people to share knowledge and experience
- Weakness in experience in cross-border transitions of businesses (e.g. Australia to the USA)
- Generally less competitive high technology manufacturing costs compared with regional neighbours
- Public sector medical research with very high quality R&D is slow to adopt an effective business culture and model
- Early seed/start-up capital difficult to assemble given the long-term return on investment

Source: Adapted from Ernst & Young 2001

deliver a range of technologies and products that will drive the global economy. Many governments have realized that they must implement public policies that facilitate/encourage further investment in the biotechnology sector, if this sector is to grow in the future. As a result, the globalization of biotechnology continues to be driven by a number of factors (Ernst & Young 2003a), including:

- Large biotechnology and pharmaceutical companies spanning global markets to sell their products and search for R&D alliance partners & collaborators;
- Multiple listings on different stock markets so that companies can expand their investor base;
- Venture capital investors seeking new opportunities with greater returns than traditional investments;
- University scientists collaborating with biotechnology companies to commercialize their research;
- Governments staking their future economic development on biotechnology innovation;
- New technology developments;
- Impact of information technology and the Internet;
- Aging population demanding improved healthcare and quality of life;
- Increased pressure on funding;
- Mergers and acquisitions;
- Collaboration and JVs;
- Convergence.

Furthermore the global biotechnology industry is being influenced by a number of discontinuities that bring about significant changes. These include the following factors:

- Human Genome Project
- Bioinformatics and pharmocogenomics
- Gene therapy and genetic modification
- Multiple analyte testing
- Biochips and multi-arrays e.g. Affymetrix®
- Nanotechnology and microelectronics
- Biosensors
- Hand-held analysers e.g. glucose and cholesterol monitoring
- Non-invasive diagnostics
- Preventive medicine e.g. vaccination
- Personalized medicine, including theranostics.

Globalization of the financial sector has become the most rapidly developing and most influential aspect of economic globalization. Compared with commodity and labour markets, the financial market has realized globalization in its true definition. This is exemplified by the number and value of daily transactions of foreign exchanges.

Although a number of indicators show that the biotechnology industry is moving towards a single global economy, it will take some time before this is achieved. What we are currently observing are examples of internationalization rather than globalization. Internationalization refers to the increasing importance of relations between nations: international trade, international treaties, alliances, protocols, etc. The basic economic units of importance are the countries and the focus remains on the economic trade between these units (Daly 2003). For example, biotechnology firms have alliances with research institutions and other firms overseas to support their business development and market entry strategies. Multinational corporations are rapidly expanding globally and are forming alliances with universities and smaller biotechnology companies. Free trade agreements have been established but between only a limited number of countries, for example between the USA and Australia.

Another observation has been that one area of biotechnology, emerging infectious diseases, occurs in areas other than the USA and Europe. For example, dengue is found mainly in tropical areas with significant outbreaks occurring in SE Asia, Latin America and Africa (Figure 13.5). Apart from a small number of biotechnology firms, these markets are neglected. The same applies for a number of other infectious diseases including malaria, Japanese encephalitis, yellow fever, and to an extent HIV, which has a comparatively high incidence in Africa.

If biotechnology promises to solve many of the world's existing problems, including infectious diseases, then why aren't more biotechnology companies focusing in these niche market segments? The main reason is that a significant market size cannot be identified by the majority of biotechnology companies. It seems that many of these problems will be resolved by smaller local biotechnology companies that become established within or in close proximity to these markets. For example, PANBIO Ltd, a Brisbane based biotechnology company currently has significant market share for its dengue diagnostic products. It also has developed a yellow fever test that has been cleared by the FDA and is developing tests for Japanese encephalitis and malaria. It can be argued that the pharmaceutical industry is well progressed towards globalization, especially when pharmaceutical companies are drawn to centres of innovation regardless of where they are located. The same could be said about western diseases such as diabetes, cancer, Alzheimer's disease, heart disease, multiple sclerosis, AIDS and arthritis. However, biotechnology will

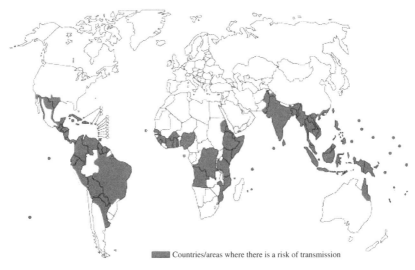

Countries/areas where there is a risk of transmission

Source: WHO, 2002

Figure 13.5   Dengue incidence 2003

not truly become global until emerging diseases in developing countries are also on the agenda of global biotechnology companies.

CLUSTERS

A cluster can be defined as a large group of firms in related industries at a particular location (Swann and Prevezer 1998). The cluster and innovation relationship has often been studied in a range of industry types, but of note in high-technology industries, is the classic example of Silicon Valley (Swann and Prevezer 1998; Kenney and Von Burg 1999). Baptista (2000) found significant differences in regional adoption and attributed these differences to clustering and network effects. Differing cluster structures, knowledge transfer patterns and factors of competition between industries can have differing innovation variables (Stevens 1997).

The facilitation and development of industry clusters, where the co-location of biotechnology firms or related industries complement each other, compete against each other or share common resources leads to increasing economies of scale (Hill and Brennan 2000).

There are five biotechnology clusters in the USA. They are known as BioBay (San Francisco, California), BioBeach (San Francisco, California),

GeneTown (Boston, Massachusetts), BioForest (Seattle, Washington; Portland, Oregon; Vancouver, British Columbia), Pharma Country (New York/New Jersey and Pennsylvania) and BioCapital (Maryland/Virginia/ Washington DC) (Biotechnology Industries Taskforce 2000). The common characteristic among clusters is the close proximity to research institutions that provide a source of ideas and developments that feed commercial industry.

Another example is the Biopolis in Singapore which is a dedicated science park providing space for lab-based R&D activities tailored to biomedical science companies. Work on Biopolis, the $300 million biomedical research hub commenced on 6 December 2001. Spread over eight hectares, the Biopolis will provide a conducive work environment for 2000 scientists and professionals in the biomedical field.

A number of clusters have evolved or have been established in various European countries to focus in the biotechnology and science sectors. The clusters of note include: Baden-Württemberg in Germany, Grenoble in France and Catalonia in Spain. Many of the biotechnology clusters and technology parks in the UK are located in close proximity to universities such as Cambridge and Oxford.

CONCLUSION

The global biotechnology industry shows enormous potential for the future. It promises to deliver a range of technologies to solve many of the world's problems. The industry is dominated by the USA and continues to grow with a significant emphasis in the pharmaceutical and medical diagnostic sectors. Indicators exist that show the biotechnology industry is progressing towards a single global economy; however, multiple opportunities currently exist to address niche markets in developing countries. This may require the establishment of small local biotechnology companies within or in close proximity to these markets. This is currently occurring in regions and countries such as India, China and the Asia-Pacific.

REFERENCES

Baptista, R. (2000), 'Do innovations diffuse faster within geographical clusters?', *International Journal of Industrial Organisation*, **18**, 515–35.
BIO (2001), 'Editor's and reporter's guide to biotechnology', report by the Biotechnology Industry Organization, USA.
Biotechnology Industries Taskforce (2000), *Biotechnology in Queensland*, Department of State Development, Queensland Government.

Biotechnology Industry Organization (BIO) (2004), 'Biotechnology industry statistics', available at http://www.bio.org/er/statistics.asp accessed 14 April 2004.

Daly, H.E. (2003), 'Globalization versus internationalization, and four economic arguments for why internationalization is a better model for world community', Available: http://www.bsos.umd.edu/socy/conference/newspapers/daly.rtf.

Deeds, D., D. DeCardis and J. Coombs (1999), 'Dynamic capabilities and new product development in high technology ventures: an empirical analysis of new biotechnology firms', *Journal of Business Venturing*, **15**, 211–29.

Ernst & Young (2001), 'Australian biotechnology report', Ernst & Young, Freehills and ISR, Commonwealth Department of Industry, Science and Resources, Canberra.

Ernst & Young (2003a), 'Beyond borders: the global biotechnology industry report', Ernst & Young.

Ernst & Young (2003b), 'Endurance: the European biotechnology report', 10th anniversary edition, Ernst & Young.

Ernst & Young (2004), 'On the threshold: the Asia-Pacific perspective', the Global Biotechnology Industry Report, Ernst & Young.

Hill, E.W. and J.F. Brennan (2000), 'A methodology for identifying the drivers of industrial clusters: the foundation of regional competitive advantage', *Economic Development Quarterly*, **14** (1), 65–96.

Kenney, M. and U. Von Burg (1999), 'Technology, entrepreneurship and path dependence: industrial clustering in Silicon Valley and Route 128', *Industrial and Corporate Change*, **8** (1), 67–86.

Mrak, K. (2000), 'Globalization: trends, challenges and opportunities for countries in transition', Vienna: United Nations, Industrial Development Organization.

PricewaterhouseCoopers (2004), 'A dose of good news', *Life Sciences Knowledge Line*, **1**, 2.

Rhodes, J. (2002), *Borderless Biotechnology*, New York: Delloite Touche Tohmatsu.

Stevens, C. (1997), 'Mapping innovation', *OECD Observer*, August/September, no.207, 16–19.

Swann, P. and M. Prevezer (1998), 'A comparison of the dynamics of industrial clustering in computing and biotechnology', *Research Policy*, **25** (7), 1139–58.

Index